RAISING CHIC
FOR BEGINNERS

The Complete Crash Course on Raising Your Own Healthy Backyard Flock to Achieve Self-sufficient

Egg and Meat Production with Safe, Smell-free Coops| Off-Grid Survival Urban Homesteading

BENJAMIN D. NELSON

This book is dedicated to my wife Erin, who supports my off-grid living style, and who rejoices on picking up the fresh eggs first thing every morning.

Table of Contents

Introduction   1

Chapter 1
A Short History of Chicken Domestication   2

Chapter 2
Chicken Terminology   6

Chapter 3
Why Raise Chicken?   9
Tips to Start and what You Need   11
what Equipment?   12
Steps for Raising Chicken in Your Backyard   12
You Can Do It   13
what Are the Challenges of Raising Chicken?   14
Chicken Noises and what they Mean   15
Dust Bath   17
Raise a Happy Flock (Facts and Myths)   18

Chapter 4
what is Your Best Breed?   20
what Nourishment Does Your Flock Need?   23
Is Chicken Raising Legal?   24
How Many Chickens Can You Keep, and what Are the Factors
to Consider?   24

Chapter 5
Chicken Feeding and Food Storage   26
Forms of Chicken Feed   27
Types of Chicken Feed   27
How Do You Store Chicken Feed?   28

Chapter 6
Know Your Chicken Coop   32
what Are the Benefits of a Coop?   33
Types of Chicken Coops   33
Traditional Chicken House   34
Custom and Homemade Coops   35
Factors for Consideration when Choosing a Chicken Coop 36
Tips for Cleaning and Maintaining Your Chicken Coop   37
Benefits of Chicken Coop Elevation   39

Chapter 7
Do You Know what Chicken Bedding Is?   44
Types of Chicken Bedding   45
How Often Should You Change Your Chicken Bedding?   47
what are the Benefits of Chicken Bedding?   47
Understand what Diatomaceous Earth Is   47
The Main Benefits of Diatomaceous Earth   48
what Could Chicken Pecking Be?   48

How Does Pecking Work in Single and Multiple
Rooster Flocks?   49
How to Stop Chicken Pecking   49
Know How to Encourage Mating   51
Common Problems During Mating   52

Chapter 8
what on Earth are Nesting Boxes?   54
Types of Nesting Boxes Available   55
Why Nesting Boxes?   56
Did You Know a Hen Can Lay Eggs Without A Rooster?   57
Reasons why Some Chickens Do Not Lay Eggs   58
Learn How to Help Your Chicken Lay Eggs
During the Cold Season   59
Types of Chicken Eggs   60
Help Your Hen Say Goodbye to Eating Eggs   61
How to Raise Baby Chickens in the Comfort of Your Home 62

Chapter 9
Chicken Health Problems   63
How Do You Know Your Chicken is Having Health Problems? 64
Ensure Your Flock Stays Healthy   65
Can Your Flock Make You Ill?   65

Chapter 10
Business Opportunity   66

Conclusion   68

Introduction

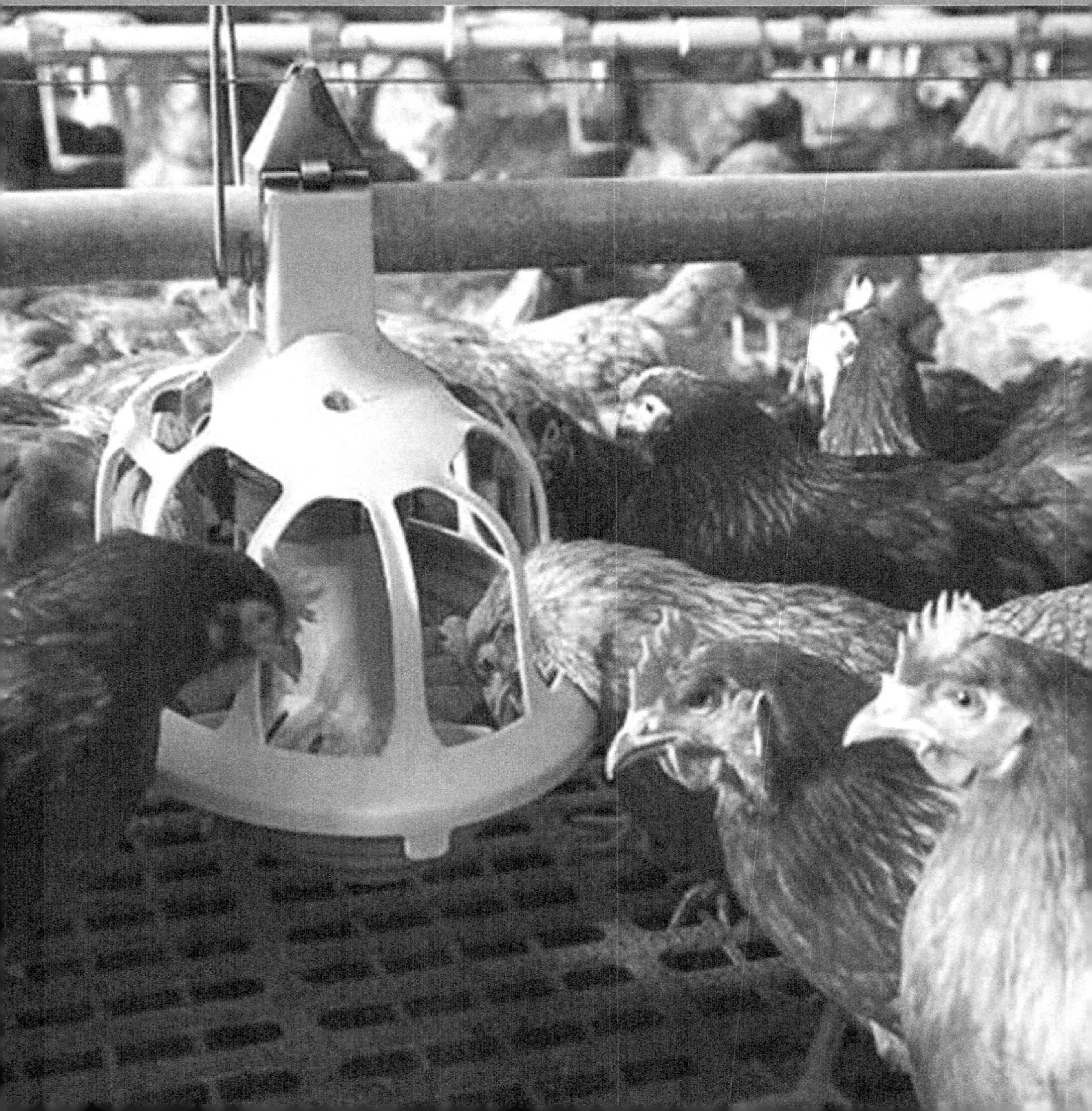

Take a deep breath and relax. You have found it! Yes, you have. A solution you've been looking for. If you have no idea what this is about, don't worry. We will enlighten you. Raising chicken! Does it ring a bell? Here is a solution for you, you could have been looking for books that will walk with you through this journey, and, likely, you haven't landed one yet. The good news is that this is the last book you will need to succeed. Your search is over. We are glad to let you know this. We have thoroughly done our research and understand raising chicken and the challenges that come with it; as you may expect, your questions have been tackled. You shouldn't be troubled about where you will start anymore. Start with this book.

Have you tried raising chicken and failed, or are you not confident enough to start, as no book has given you what you want? Or is it that you have no idea about raising chicken? We don't give empty promises. We deliver, so don't put this book back on the shelves. You will miss a lot. Have you been looking for a book that will educate you about everything you need to know about raising chicken, from how you do your research to purchasing what you require, for example, the best chicken breed and coops in the market, to enjoying bonding with your happy and healthy flock? Worry no more! We've got you. If you are new here, we have made it simple for you. New and interesting things for you to learn, and it won't be surprising if you end up as the best chicken keeper in your area. How great! Our book has everything that will make this intimidating project fun and successful. You can start planning your journey as it is about to begin. Are you ready? Please come with us.

Chapter 1
A Short History of Chicken Domestication

To predict the future, you should understand the past. The past will help you understand how things were originally and the best way to do them. It provides solutions for modern-day domestication. You can determine the best coop to use for your chicken or the best way to feed your flock. Raising chickens dates back 7,000 to 10,000 years. It is known to have occurred around 2000 BC in Indus Valley, although the West and Zhou propose an earlier origin around 6000 BC in Southeast Asia. Based on archaeological evidence, chickens were originally used as crow cocks to welcome the new dawn and, in a few years, used for cock-fighting and pet.

Cock-fighting was discovered by the Athenian general known as Themistocles when he headed for the invading Persian forces. Themistocles stood and watched as the cocks fought, and he was impressed by it. He summoned his troops and told them that the cocks didn't fight for glory, household gods, or liberty but rather because they couldn't give way to each other. The troop found this inspirational, unlike what you might have expected, pointless and depressing. The Greeks continued to repel the invaders. There is no account of what happened to the loser cock in the tale, but you sure know that this was the beginning of cock-fight, thanks to the winning cock, which is now honored by being fried, breaded, and dipped in your favorite sauce. Selective mating aided the rise of commercial chicken over the last 100 years. Many scientists agree that the primary ancestor of the chickens is the famous Southeast Asian Red Junglefowl, also known as the gallus. This chicken lacked the yellow skin gene; therefore, it is believed to be hybridized with the gallus sonneratii of India, also known as the Grey Junglefowl. Indigenous chicken acquired varying genetic characteristics and aided adaptation to different environments and conditions such as humidity, heat, and diseases.

Chicken domestication grew fast as it provided meat and eggs without competing for human food sources. Today, crossbreeding between indigenous and commercial chickens is done to ensure the indigenous traits carry on, although the traits decrease with repeated crossbreeding. The use of mitochondrial DNA conducted the study of the genetic diversity of the chicken. The association between a junglefowl and a commercial chicken may aid in understanding the geographical origin of domestication. It's key to note that chicken is a sacred animal in some cultures in the world. It contributes to culture, art, and religion. For example, an ever-watchful hen was seen as a symbol of fertility. The Egyptians hung eggs in their sacred places to ensure that what they called their bountiful river remained flooded. For the Romans, it was further interesting. The chickens accompanied the soldiers in battle. How the chicken behaved determined how well or badly the battle would go. If the chicken had a good appetite, it meant that victory was for them, and you can imagine how it went when they watched their chicken refuse to eat.

The United States Chicken Council has been in existence for many years. It is a non-profit organization that champions the United States broiler chicken interest. This council dates back to 1954. It was then known as the National Broiler Council, which changed to National Chicken Council in 1998. This council is an important source in understanding the history of domestication. According to this council, poultry farming existed in the 1800s and early 1990s, but it was mostly yard farming. This was aimed at providing food for the family. The chicken was mostly for the eggs, and the meat was for special occasions. When a chicken was slaughtered, people made sure they were around to enjoy the dish. Although chickens were mostly for their eggs, the farmers did not understand how important sunlight was for egg laying, and therefore the rate of production was low.

During the 1900s, it became more interesting. Poultry farming became a business opportunity. These farmers sold chicken and eggs to people who didn't view raising chickens as a means of making money but rather as a dish. As years went by, around the 1920s to 1930s, the farmers understood the basics and benefits of having a healthy flock. They invested in ensuring the flock had enough space to move freely, clean and healthy water for the flock, and good food. Weather was also critical. The farmers had to seek a way to shelter their flock from harsh conditions such as rain. Rain doesn't work well with chicken raising as it causes wet and muddy conditions, which make the chicken sick. During this period, egg production had grown compared to meat production.

For this reason, the farmers turned to broiler farming, thanks to Mrs. Wilmer Steele. She kept 500 chickens which she sold for meat. After some years, in the 1940s and 1960s, all industries had to work together, consisting of farmers, hatcheries, and feed millers. To start chicken farming, you had to get chicks from the hatcheries. This was because egg production was technical, and some of these farmers wanted to avoid that. Feed millers also came in handy they provided necessary feeds for the farmers to purchase for their flock. If a farmer couldn't afford the feeds or chicks, they offered loans, which was a crucial industry back then. what more could the farmers ask for? The market grew, and the United States exported broiler chicken to the Soviet Union. This demand in Russia and other countries went up, which meant that the level of production in the United States hiked too; hence consumer protection had to be implemented, and they came up with a Hazard Analysis and Critical Control Point in all the large-scale slaughterhouses. The more the chicken industry developed, the more the chicken decreased in cost. With all the research and industries dedicated to helping farmers in the world, it has become easier to domesticate a chicken, you understand the challenges different farmers have faced, and how they overcame these challenges, it's no longer as challenging as before. Why don't you try it today?

QUICK CHAPTER TAKEAWAY

You could think there isn't much history of chicken and chicken domestication; chickens have existed for many years. On the contrary, there are loads of interesting things to learn. Can you imagine a chicken going to battle or a single chicken fight turning into something amazing? You should make an effort to understand what a chicken's journey has been like to be able to properly take care of and understand it. Understanding the key persons and industries in chicken keeping is also important.

Chapter 2
Chicken Terminology

If you are new to chicken-raising, some terms could be challenging. You might know something about chicken raising if it's not a new term. It doesn't matter the situation you are in, we understand you, and that's why we've crafted this chapter to ensure you have a smooth ride in the other chapters. Welcome and walk with us!

- **Chick** is a young bird, mainly one newly hatched (baby chicken).
- **Cock** is an adult male domestic fowl.
- **Hen**, this is a domesticated junglefowl species with some characteristics of the wild Ceylon and grey junglefowl.
- **A rooster** is an old chicken that weighs about 5 to 7 pounds and is usually about 3 to 5 months old.
- **A broiler** is a chicken raised to provide meat. It is a hybrid chicken. Hybrid enables fast growth and finish. A broiler chicken weighs about 5 pounds and is white.
- **A feeder** is equipment used by farmers to feed their flock. You put chicken food in the feeder, and the chicken is fed from it. Your number of birds determines how many feeders you will buy or use. For the safety of your flock, you should ensure that the feeder is not too crowded and clean.
- **A nesting box** is an enclosure where chickens nest.
- **Bedding** is a farmer's material to spread around the coop and the nesting box. Unlike litter, bedding is new and clean.
- **A coop** is a structure where chickens live in.
- **Run** is an outdoor fenced-off area attached to a coop where chickens move freely during the day. A run has two important purposes: keeping the flock safe inside and keeping predators away.
- **Predator.** This is an animal that naturally preys on the other. It kills and feeds on it. Chickens are more prone to predation, unlike the turkey, due to their size. The chicken predators include but are not limited to dogs, raccoons, foxes, coyotes, hawks, wolves, red foxes, bears and snakes.
- **Chicken wire** is a mesh made of a thin, flexible and galvanized steel wire and is used in a coop or a run keeping the chickens in. Its durability lasts more than five years, although it could be affected by moist soil or wet climates.
- **A brooder** is a heated house containing an infrared lamp and a hood that directs heat towards the chicks' floor. Its purpose is to keep the chicks warm as cold seasons affect chicken raising.
-
- **Molting** is replacing old, all or some chicken or bird feathers with new ones. Molting occurs annually, and it lasts from 8

to 12 weeks. Some molting symptoms include loss of feathers and no or reduced egg laying. The following tips guide you and help your flock get to this stage easily. Feed your flock with proteins during this period. The chickens require protein for feather regrowth. Just like humans, chickens do not want stress during molting. This is like a vacation for them. They need to relax without a lot going on and provide an area where they can move freely, as being in a coop is uncomfortable. Ensure you handle them with care and provide a clean environment with clean and healthy food and water. Once the molting period is over, feed them what they need at the moment. In simple words, you should meet their nutritional requirements.

- **Cockerel,** this is a young domestic cock.
- **The hatchery** is an industry where eggs are hatched under artificial conditions. The aim is to hatch the eggs and then sell them to the farms, as many farmers do not want to deal with this technical part.
- **A pullet** is a female baby chicken, mainly one that is less than a year old and hasn't started laying eggs but is about to. The price of a pullet is higher than that of a chick as the incubator has invested in feeding and keeping it safe.
- **Layer Breed**, these are highbred chickens that require raising from when they are one year of age. They start laying eggs at week eighteen and continue until 72 to 78 weeks. The most commonly known layer breed is the White Leghorns.
- **Dual Purpose Breeds** are chickens that are productive at two or more traits. This means that they are raised on both eggs and meat. Some duo purpose breeds include Light Sussex hen, Orpingtons and Plymouth Rocks.
- **Waterer** is a container that holds and delivers water for chickens. They are basic equipment made of either metal or plastic. Your chicken water should not be easy to topple or too close to the bedding as it will contaminate the water.
- **Dust bath**, bathing in dirt or other substances to help remove external parasites and groom plumage.
- **All flock feed** is a type of feed that meets the nutritional needs of a mixed flock. This feed contains vitamin D, which aids in hard-shelled eggs, natural plant extract and essential oils. This helps consumption growth and bone formation.
- **New castle disease** is a highly contagious poultry and bird disease. It is caused by a virulent strain of the disease virus.
- **Juvenile,** a young male or female bird.
- **A chicken tractor** is a mobile coop without a floor.
- **Ventilation** is the intentional provision of air in a room and, in our case, your chicken coop. It aims to provide indoor air quality by diluting pollutants in the coop.

Chapter 3
Why Raise Chicken?

You may question why you have to raise chicken when you can get all the chicken products from a nearby supermarket. Furthermore, chicken-raising requires a lot of effort and energy. While it is true that you can walk to your favorite chicken place and get your products in minutes, do you want to give up the fun and experience of raising your chicken at your home? The benefits of raising chicken are unmatched. Once you start your journey, you'll not turn back. We are here to make sure that doesn't happen. We have prepared the information you need to help you understand the wonderful benefits of raising chicken. If you want to start this project, the information will help you. If you are in the middle of all this and feel like it is not what you hoped it would be, we are here for you.

EDUCATION

How does chicken-raising help with education? You may ask yourself. Have you tried to ask a child to name the parts of a chicken, and they can't do it? Or rather the origin of a certain species? These children are used to ready-made chicken and don't even know where it comes from or its biology. This ends once you start raising your chicken. Your child will be able to witness how a simple egg transforms into a full-grown chicken as well as easily name all the parts of a chicken. The child will also be involved in all the activities such as feeding, treatment and cleaning. These are very important life skills. Chicken-raising will help you understand the history, business, legal issues, food safety, math and genetics. what an easy way for your child to learn. Isn't it great watching your child learn what it takes to put food on the table?

SOURCE OF INCOME

If you start chicken raising as a hobby, it might be a business opportunity. You may also start it as a business opportunity. It will get to a point where you cannot eat all the eggs alone. You end up selling them and generating income. The same applies to chicks and chickens; you could sell chicks to people who want to start a business or even sell roasters for holidays and events. Selling feathers and manure is another business opportunity if you have no plan to use it or have more than enough.

ENTERTAINMENT

Watching your flock in your backyard can be therapeutic. Often people with chickens get a seat and enjoy the view of the flock moving around, say bye-bye to those boring and lonely days, and have a chicken for a pet. Did you know that just like any other pet, chickens are affectionate? You'll see them jump on your lap so you can pet them and bond. Most time, this pet will welcome you home. They'll run to you to see what you brought them. Your child could also play with the chicken. Imagine your children playing around with them. Great thought! It's never a dull moment with a flock in your backyard.

FOOD QUALITY

Knowing where your food comes from and controlling the quality gives you satisfaction and comfort. When feeding your flock, you understand the quality you'll get in the final product. You don't have to worry about antibiotics or any other additives. Chicken raising is also a source of food. You get loads of eggs which you can use to bake or make as you like, chicken meat too! Anytime that chicken curry craving hits you, you can slaughter one of your flocks. You can never go hungry with a flock in your backyard.

WASTE REDUCTION

Chickens eat a variety of foods. Therefore you don't have to stress over leftovers. Although lots of leftovers are not recommended, you do not have to worry about the 10 percent left during breakfast or dinner. Your flock got you. The more your flock feeds, the less you have to dispose of. Isn't that amazing?

NATURAL INSECT CONTROL

Chickens have an instinct to scratch around and pick anything they can eat. Apart from what we all know, weeds and plants, what else could they probably feed on? Bugs! Yes. A chicken will eat any bug that it comes across. This includes things like spiders, ticks, ants and grasshoppers. It's also crucial to note that because the chicken eats all bugs, it has no effects on it. Indeed, your flock may fall sick, so make sure you are keen on that. Worried about bugs in your backyard? Your flock will aid in bringing the number of bugs down.

FERTILIZER

People around the world purchase chicken manure. The key word is 'purchase.' With your flock, you don't. Your flock will give you the manure. They will make a nutrient-dense fertilizer. Although it's not advisable to use fresh chicken manure as it will burn your plants, you can pile it in your backyard compost bin, leave it to boost nutrients, and then use it for your garden. Double benefit!

HEALTH BENEFITS

Taking care of living things boosts mental health. Taking care of your flower garden lightens your mood and enables you to enjoy the fresh air. When you are having a difficult day, you can spend time with your flock. It will be a stress reliever and a mood booster for you. Let us not forget the physical benefits. Having a flock means you have a daily routine. You have to feed them, clean their coop, change their drinking water, check if all of them are healthy, fix broken things, and collect eggs. These activities will aid in keeping you fit. It doesn't have to be so complicated.

YOU BELONG

If you are doing it, someone else has done it. Another chicken farmer is out there. Your friend might be doing it, or even your co-worker. Great! This means you have a group you can connect with. You could join seminars, make new friends, and learn and share ideas. Being around like-minded people with similar interests or hobbies will make your journey easy and interesting. Don't be left behind.

Tips to Start and what You Need

For you to start your chicken raising project, there is the basis that you need to understand to make the journey easier and also ensure you get yourself the necessary equipment to avoid unnecessary challenges. The following are tips for ensuring a smooth ride.

- Determine whether your local authorities permit chicken raising. It will be frustrating to invest time, energy and money only to realize that you cannot keep your flock. Some authorities will not permit chicken raising while others will allow and others limit the number of chickens you keep. Ensure you understand what your local authorities say about chicken projects.
- Choose the right chicken breed; there are different breeds, so you should research and pick one that suits your preferences. If by any chance you know someone in your neighborhood who does chicken farming, you could ask for advice on the breed that does well in the area.
- Decide the maximum number of chickens you want to keep. What is the size of your backyard? What does the authorities say, and what size of a coop do you want? You shouldn't overcrowd chicken as it might risk the health of all of them, so ensure you get your math right.
- Do your research on a reputable chick supplier. The last thing you want is to make a deal with an unreliable chick supplier. Things could go wrong right from the beginning. If your research requires a lot of time and effort, do it. You would rather spend a lot of time on your research and get it right rather than messing up everything.
- Determine the cost of the project. The budget is very important. Calculate your budget before bringing your chicks home or building that beautiful coop. You do not want your project to get stuck in the middle. This project requires money. Your chicks need to be fed and looked after.
- Pick the right size and design for your coop; if you want to start with adult hens, your coop should be different from that of chicks, and you should invest in fencing and overhead netting. You could go for a homemade coop and make it how you like it or buy a commercial one. It is easy and everything is set up correctly.

- Protect your flock against predators and harsh climate. It doesn't matter the area you are in. Ensure your chicken coop has roof coverage.
- Choose a quality brooder and know where to put it. Your chicks need a brooder before anything else, and it should be indoors, preferably in a garage.
- Ensure you have a feeder before bringing your chicken home. Feeders ensure that the flock eats properly and minimizes feed wastage.

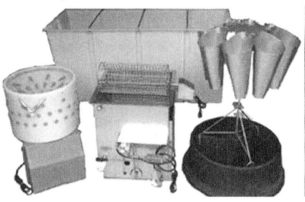 

What Equipment?

- A coop to keep your flock safe during the night.
- A run if you are not comfortable with free-range.
- Mirrors and chicken swings if your flock is in a run to save them from boredom.
- Chicken feed.
- Feeders to protect your flock's food from scavengers and wastage.
- Waterer to ensure your flock remains hydrated.
- Brooder and incubator, it's important to understand how your chicks behave in a brooder to meet their needs. If you see the chicks form a ring around the lamp, it's too warm in the brooder, and if they move closer to or under the lamp, the brooder is not warm enough.

Steps for Raising Chicken in Your Backyard

Plan the size of the brooder you need. You first need to know that when your chicks arrive home, they need a safe place to be. A brooder box will do the magic. If you are raising chickens as a hobby, the size of four chicks will work for you. If you want a bigger number, then plan for a bigger one.

Determine whether you have enough space. Do a survey and determine if you have enough space to raise your chicken. Can

the structure you want fit, and do the authorities agree with the sizing and building of chicken shelters?
Go legal. Put together all your site and coop plans and head to the local authorities, have all the required documents and ensure that you are cleared to start your project. You do not want to have any problems after you've started.

Get your flock a coop. The coop should fit a brood properly; you can purchase a ready-made one or do your research and build one that suits your needs. Whether you go for a commercial or a homemade one doesn't matter. The size should be big enough and durable.

Install your flock home. After getting your chicken coop, it's time to install it. If you purchased one, ensure you follow the manual to assemble it correctly. If it is a homemade one, ensure you do it correctly, and if you need any help from your neighbors, don't be shy to ask. It might be a fun activity for them.

Let's get the supplies. Once the home is ready, get some supplies for your chicken. Your chicks will need a heat lamp in the brooder, bedding and nesting boxes. Do your research to determine what type of feed you need for your flock.
Set up the nesting boxes and the brooder, as you did with the coop, and follow the manufacturer's instructions to assemble. Ensure you have a heat lamp set up.

Assemble and test the feeding equipment. It will be disappointing to bring your chicken home only to realize that the equipment isn't working properly. Please give it a test.

Get your chicken, the long-awaited moment. You may have done your research and have a reliable supplier. If not, spend some time doing your research and consulting knowledgeable people. Get to know the right breed to bring home.

Keep your chicken home clean, ensure the coop and run always stay clean and check for eggs daily. Have a cleaning schedule.

You Can Do It

Anyone can raise a chicken as long as you have the requirements. Chickens are easy to keep and give us more than one product. You might wonder how it would go after you take a step to start. You could also be worried about disappointing your family, especially your kids, who are happy to see chicken around the home. Your worry could be that you don't want your kids to see you fail. Relax! Our book is here for you. If you have no experience, don't worry, we've got you.

Keeping chicken will be a great experience for you. You can wave goodbye to the worry of not knowing where to start from, the type of coop you need or the breed that will do well in your area. Our book explains all this, so choosing what you need is in your hands. The first few months could be challenging as you have to get used to the new life, but we will walk with you. Chicken raising is a rewarding experience. You'll not regret taking this road. You need to understand simple things to get the desired result. The following are some of the basics to make it work.

- To understand how to take care of your flock, you could consult knowledgeable people on the subject matter.
- Realize that challenges come with chicken-raising and prepare how to tackle them.
- Believe in yourself. You might feel unprepared and start questioning the type of setup or breed you have. This doesn't help. Be confident with what you have done, and you will get positive results.
- Invest in learning how to bond with your flock. Just like human beings, animals need love and to be cared for. If your flock is happy, it will make you happy as well.
- Get support from your family and share the experience with your kids and close friends. Educate them, and let them educate you. In case of any concerns, let your kids know what is happening.

What Are the Challenges of Raising Chicken?

- For zoning, check with your local authorities. It can be heartbreaking wanting to start a chicken project, but you can't because your local authorities won't let you. If you are in urban areas, roosters might not be allowed.
- Deciding on the best home for the flock and choosing the right coop will contribute to its success. Having the wrong coop could result in your flock having behavioral issues such as egg eating, predation, feather picking and general frustration.

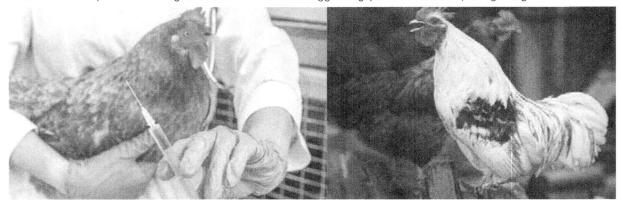

- Chicken diseases. Your flock could fall sick, and some may die. Chickens will fall sick if exposed to germs, bacteria, viruses, fungi, and external and internal parasites. Be on the watch to make sure your flock is healthy.
- Predators. Without proper housing or lifestyle, your chicken will be exposed to predators.
- Deciding the best lifestyle for your chicken is a challenging task. Each lifestyle comes with its challenges. If you decide to settle for free-range, you are risking predation. If you settle for confinement, your flock is at risk of suffering from

behavioral problems due to boredom. There are other health issues such as obesity and foot pad infections. Weigh the advantages and disadvantages of each lifestyle and pick the better one.

- Boredom could lead to undesirable character. Your chicken could have nasty habits such as self-mutilation, egg eating and bullying. As mentioned earlier, you could get a mirror to help reduce boredom. It's one of the many solutions.
- Noise. I would be lying if I told you that chickens are quiet and that you can be sure not to get any disturbances from them. Chickens are noisy, especially if it's a rooster. It's advisable to talk to your neighbors if you are planning to start this project. It's also important to note that some authorities allow raising hens but ban roosters due to too much noise. Hens can be noisy too, especially when they are in danger or just laid an egg; you should get to learn the different noises that your flock makes.
- Smell. The chickens are very clean as they dust bath from time to time, but the coop could end up smelling due to their waste. Chicken poop smells, and therefore, it is important to clean up once in a while. If you do not clean the coop, your whole house might smell of chicken poop.
- Time. Chickens are low maintenance, but you still invest your time in ensuring that they are cared for. That cleaning or feeding takes time.
- Cost. Chicken raising requires you to invest money to buy the coop, feed and water you need. Caring for your flock requires money. Medication is money too. Have a plan.

Chicken Noises and what they Mean

Amazingly, chickens can make over twenty-four noises and even stitch these noises together while having a 'dialogue' with other chickens. Do you have any idea what these noises mean? The noises go from food to danger. By choosing to raise chicken, you should understand what these noises mean. Allow us to help you.

COOP CHATTER

You will hear these noises immediately after you free them in the morning. They could make a happy coop chatter as they fled to enjoy the day freely or make complaining sounds, especially if you are late to open the coop door. This chatter also happens when you bring them back in the evening, they have had a good day, and they are ready to relax for the evening. By this time, they murmur and make contented sounds.

EGG SONG

It's possible that you already know this one. Hen makes happy cackling sounds after laying an egg. The other hens might join in the song, especially when they are close. The sound could go for a while, especially if a number of them lay simultaneously.

CONTENT MURMURS

Most times, when chickens are moving around the backyard, they make this sound. You'll either see them together or hear these low murmurs. No reason to worry if you hear these sounds. Your flock is safe. In case of any danger, they will alert each other. They know how.

BROODY HENS AND GRUMBLES

A hen will make this sound when it's sitting on the laid eggs, waiting to hatch. It will growl if the hen feels threatened by you or another chicken. This sound warns you to stay away. If you don't, it's going to peck until you leave. Some hens will even scream at you, fluff up and give you an evil look. During this time, the hen leaves its nest once a day and is in a bad mood constantly.

ALARM SOUNDS

These birds have a sophisticated range of dangerous sounds; there are those alarm sounds for ground predators and aerial predators. It's important to differentiate the two to protect your flock properly. The chickens know that as long as you are near them, no danger will befall them; therefore, they know when they make the sound, you will come to rescue.

FOOD NOISES

When a hen calls its chicks to eat, you'll hear the tuk tuk sound. This sounds alerts the chicks of a nice meal. The mother will try her best to ensure the chick has eaten. The tuk tuk sound is a sign of pleasure, and it's not used often, only when there is a treat.

MOTHER AND CHICK SOUNDS

Communication between a hen and its chick starts before the eggs are hatched. You will hear your hen cluck and purr when moving around or sitting on the eggs. The mother does this so that the baby can differentiate its mother's voice from the rest of the flock. When the time is almost due to hatch, you will hear the chick talking with their mother, and she will make them feel safe and encourage them to break the shell. Like human beings, when a chick takes a long time to learn, the hen will slow down to make sure it does. In case of danger, the mother will make the 'grrrr' sound, and the chicks will run for safety.

QUIET CHICKEN

All chickens make sounds, although some make more noises than others. If the noisy chicken goes silent, it means something is wrong. Maybe it feels isolated or just having a bad day. Chickens gets depressed. Therefore, keeping an eye on your chicken until it returns to normal is recommended.

You could have thought chickens didn't have a language all along. Science proves this wrong. They do. Your new project will

help you understand this. Just sit at a distance and watch how your flock interacts and pick out the different sounds they make. It will for sure be an interesting experience. We have listed the main sounds. Time for you to learn!

Dust Bath

Watching your flock dust bath could worry you as you might think your chicken is getting dirty. This is the opposite. Dust bathing is enjoyable and leaves your flock clean. This bath also helps them keep off parasites such as lice. Most of the time,

you will find almost all of your chickens doing it as they like doing it in groups. This is a great time for them to socialize. Please don't stop them; instead, help them. Parasites are so annoying, and they need to get rid of them. This is how you help.

- Dig a hole or make a place for them to dust bath. Make sure it's able to hold the bust bath mixture.
- Put enough sand or fine dirt; it will be easy to bathe and get in the feathers.
- Add some diatomaceous earth. Adding diatomaceous earth is a great remedy against parasites such as lice and mice. This is a natural insecticide; therefore, it's safe for your flock. A handful of diatomaceous earth is enough for a bath.
- Add herbs. Fragrant herbs make your flock smell nice and also have natural insecticide that kills parasites. These herbs include rosemary, mint and lavender.
- Make sure the bath is covered, the bath shouldn't be left open, and the dust will work better when dry. Find means that will make sure the bath remains dry. A tent or an umbrella will do.

Raise a Happy Flock (Facts and Myths)

Chickens need to be happy. If you treat your flock well, they will be good to you. If you want to start a chicken business treating your flock well is the secret. Here are some ways to ensure your flock stays healthy and happy.

AVOID KEEPING A SOLITARY CHICKEN

No animal wants to be isolated, so it's important to ensure that your chicken has company. Chickens like to socialize, and you will even see them taking a bath together. If you are raising chicken as a hobby, you could keep one or two. This is not advisable because raising one chicken means it's isolated; if you keep two of them in case of predation, one will be left lonely. To avoid this, ensure you have three or more chickens.

BALANCE DIET

Like human beings, chickens need a balanced diet. Chickens are not picky and might end up eating anything that comes their way, but this shouldn't mean that you should neglect the nutritional needs of your flock. Buy formulated feed for your flock and give them the right portions. A balanced diet will ensure your flock stays healthy and productive. If you want your flock to serve you well, serve it well.

GIVE ROOM FOR EXERCISE

As mentioned earlier, chickens will have health issues if not allowed to exercise. Allowing your flock to exercise will make it bond and love you as you are concerned about its health. Exercise makes your chicken a happy chicken.

MEDICAL CARE

If you don't pay close attention to your flock, you may not realize when one is sick. Therefore, it would be clever if you inspected it from time to time. If there are any sick chickens, ensure they get medical attention as soon as possible. Most chicks from hatcheries are usually vaccinated. If you have any concerns, consult your supplier and find a solution. A healthy chicken is indeed a happy chicken.

SPEND SOME TIME WITH YOUR CHICKENS

Just as you may spend time with your cat, spare some time to bond with your flock. You may have noticed that someone might be raising chickens, but anytime the chicken sees them, they flee. It's because the flock hasn't bonded with you and they are scared. Bonding with your flock doesn't require too much work. You could carry a chair that's not too high and sit close to them or in their coop. If the flock doesn't come closer to you, you could decide to sprinkle some food near you and let them come to you. You don't need to keep looking after them or cuddling them. You could continue doing what you like reading a book or watching the news on your phone. The flock will eventually feel safe and happy and might start pestering you. Be careful not to send the flock away during this time.

KEEP THE FLOCK ENTERTAINED.

Entertainment is key in chicken raising. Allow your flock to move around your backyard and let it do what it likes. If your chicken wants to dust bath or peck, let it do it. You could also install mirrors in the run to limit boredom. Animals get depressed. It would be wise if you didn't let your flock be a victim.

AVOID OVERCROWDING

It's impossible for your chickens to be happy in a crowded area. When choosing a coop, estimate the chicken you want to raise. If you want more chickens, expand the chicken coop. It would help if you let your chicken be chicken. Let the flock have enough space to scratch and flap happily.

DO A PARASITE TREATMENT REGULARLY.

A parasite stays on or in a host and gets its food from or at the expense of its host. How can your flock be happy with parasites? You should do an inspection and treat the coop occasionally. If your flock is affected by coccidiosis, you should feed your flock with medicated feed and water.

Here are some of the chicken facts that you should know:
- They are omnivores.
- Chickens have better color vision than humans.
- Chickens do dream.
- Chickens communicate, and every vocalization has a meaning.
- The egg color is determined by genetics.
- They evolved from dinosaurs, and therefore they are technically dinosaurs.
- Chickens are smart.
- Some chicken breeds are on the verge of extinction

MYTHS

- For your chicken to grow faster, you should give them hormones and steroids.
- Chickens are genetically modified.
- Broilers are raised in cages.
- If you give your flock antibiotics, they will remain in the chicken meat.
- Plant-based meats are better and healthier than chicken meat.
- Chicken raising and production are not sustainable.
- Chickens grow so fast because they are so huge that they can barely stand up.

QUICK CHAPTER TAKEAWAY

It is important to understand why you need to raise chickens before you do anything else; start this project with a purpose that will be your driving force. If you raise chickens just to raise them, your project could get stuck in the middle. After helping you with this, it has been our pleasure to reveal the 'secret' or, let us say, the tips, equipment, steps and challenges of chicken-raising. Stay with us as you'll get to learn more.

You should answer some questions before deciding on what breed you want to keep. You do not want to get stuck in the middle. The following are questions to answer before anything else, why do you want to keep chickens? Is it because of eggs, hobbies, pets or meat? How large is your backyard, and can it fit the number of flocks you want to keep? what color of eggs do you prefer, and after how long do you want your flock to start laying eggs? what is the climate in your area, and what coop design do you prefer? Get these questions answered, and you will be ready to start your chicken-raising journey.

AUSTRALORP

This breed originated from Australia. Australorp is their national bird. It has three varieties blue, black and white, but it is mostly recognized in black. This breed lays the highest number of eggs in a year, 250. This breed starts laying at 16 weeks and is a good mother. They can also be kept in different weather conditions and still survive. They are also a great choice for a pet, especially pets for kids.

BARNEVELD

These chickens are also known as 'Barneys.' This is to show affection. They are low maintenance, attractive and have a great personality. They have beautiful feathers with a beautiful patterning simple but amazing design. This bird is rectangular and has high wings on its body, which makes it impossible to fly. This breed is known for its eggs. It lays about 200 eggs a year, which means it will bless you with three to four large eggs in a week. A Barneveld lays dark chocolate eggs, which are occasionally speckled. They still lay eggs in winter. They are also known to be easy going, and therefore are a good choice for a pet. An interesting fact about these birds is that they like being set free and will excel in it.

BRAHMA

This is an American breed of chicken. Brahmas are massive in size. The only breed that can rival a Brahma is a Jersey Giant. Although they are intimidating, they are gentle birds and protect the rest of the flock from small predators. Brahmas prefer laying eggs over the cold seasons. They thrive better in brutally cold areas such as the northern states. These birds come in dark, buff and light colors. The disadvantages of raising these birds are that it might take seven months to start laying eggs, they eat lots of food and feeding them can be expensive. When they are hungry, they can bully the rest of the chickens.

BUCKEYE

This is the only breed that was created entirely by a woman. She is famously known as Mrs. Nettie Metcalf. It is a dual-purpose breed and has yellow skin. Other characteristics include a red pea comb, cold weather handy, a red face and earlobes. Although this breed does well in almost all living conditions. A buckeye will not allow rats to roam around. They are excellent hunters, and mice won't be a problem in your backyard anymore. This breed also does well during cold seasons, and it will lay about 200 eggs in a year. These birds love attention and therefore are a great choice for a pet.

JAVA

Java is an American breed. These birds were the second ones to be developed in the United States. They are heavy birds with distinguishable bodies. They are beautiful birds with many uses and can easily adapt to the environment. They are dual-purpose birds. They produce eggs and meat. These birds are slow growing. Unlike other birds, it takes about three to four months for a Java to reach the market size of six to eight pounds. There are different varieties of Java: white, mottled, black and auburn. The common ones are the black and mottled Javas.

JERSEY GIANT

These are the largest of all chickens. They are around two feet tall and weigh around 15lbs. They are an American breed and are rarely found, which shouldn't be the case. Although they are big and can be intimidating, they are very friendly birds and get along well with humans and the rest of the flock. They do well in all lifestyles, from free-ranging to backyard runs. Their size helps them produce a good amount of meat and a good number of eggs.

NAKED NECK

This is a breed that naturally has no feathers on the neck. It is also known as the Transylvanian Naked Neck or a Turken. This bird is known as a Turken, which is believed to have been a hybrid of a domestic turkey and a chicken. This breed is commonly found in Europe and South America. It has a dominant trait controlled by one gene and is easy to introduce to the other breeds. The skin is yellow in color, light brown egg and a single comb. Some of the characteristics of this breed include duo-purpose utility, immunity to most diseases, a single comb, and few feathers. When it's too hot, the head and neck turn red.

ORPINGTON

Orpington is a British breed. They are large with a single upright comb. These birds are fun to raise and are very friendly. They are found in various colors: chocolate, lemon cuckoo, blue and diamond jubilee. They are big and fluffy and love attention. It is a good choice if you have children as they love to be held. They also have a better layer of eggs, make great mothers and have a docile personality. An Orpington loves to free-range and look for bugs; therefore, you don't have to worry about bugs in your backyard with this breed. Although this is the case, these birds can spend time confined as long as they have enough space to move around. Surprisingly this breed would prefer to spend time sitting around the feeder enjoying the food. They don't work for the food as long as it is available.

PLYMOUTH ROCK

As its name may suggest, the breed originates from Plymouth Rock. A few varieties are accepted in the United States, including buff, blue, barred, Colombian, partridge, white and silver. This breed is best suitable for first-time flock keepers and seasonal flock keepers. The breed is known for being calm, responsive and highly relaxed. If you have a Rock in your coop, be sure that it will remain lively and active as this breed has a fun personality. They also lay large brown eggs and have an average of 200 eggs per year. Their egg productivity does well in the first three years and declines afterward.

RHODE ISLAND RED

The Rhode is an American breed. They are famously known for their egg-laying abilities, attractive feathers and delicious meat. You could choose this breed if you want a small flock or a pet. This breed has a large and heavy body with red coat feathers and a shining black tail. Although this is a dual-purpose breed, they are commonly known for their egg-laying abilities, 200-300 eggs per year. They also take longer to reach the recommended butcher size, say four to five months. Amazingly a Rhode adapts to any environment easily, can still survive in harsh weather conditions and with a poor diet, and it will continue laying eggs.

SUSSEX

Sussex is a British breed of dual purpose. They are friendly birds. Sussex is the way to go if you are looking for a friendly flock. You will never feel lonely with these birds. The oldest and most common variety is the Speckled Sussex. This bird dates back to a hundred years ago, and now it is still a favorite for many. These are some of the factors why you should choose Sussex. They are gracious mothers. They are kind and curious birds and lay a good number of eggs. 200 to 250 brown eggs in a year.

Here are some of the breeds that will make the best pets
- Brahmas
- Cochin
- Jersey giant
- Orpington
- Rhode Island Red
- Wyandottes
- Silkies and Speckled Sussex

MEAT CONSUMPTION
- Sussex
- Wyandotte
- Orpington
- Turkey
- Jersey Giant
- Dorking
- Delaware and Buckeye

LAYERS
- Sussex
- Australorp
- Barnevelder
- Brown Leghorn
- Buckeye
- Buff Orpington
- Wyandotte
- Well summer
- Rhode Island Red
- Plymouth Rock
- New Hampshire

ALL–AROUND CHOICE
- Marans
- Brahma
- Delaware
- Wyandotte
- Turkey
- Sussex
- Plymouth Rock
- Orpington
- Egyptian Faiyumi

WHITE EGGS
- Minorca
- Ancona
- Leghorn
- Dorking
- Andalusian
- Buff Sablepoot

BLUE EGGS
- Easter Egger
- Araucana
- Ameraucana

GREEN EGGS
- Oliver Egger
- Isbar
- Easter Egger
- Favaucana

BROWN EGGS
- Orpington
- Australorp
- Welsummer
- Turkey
- Sussex
- Plymouth Rock
- Rhodes Island Red
- Marans
- Jersey Giant
- Delaware
- Buckeye
- Braham

DOES BEST IN COLD CLIMATES

- Australorp
- Brahma
- Barnevelder
- Buckeye
- Wyandotte
- Welsummer
- Rhode Island Red
- Plymouth Rock
- New Hampshire Red
- Jersey Giant
- Faverolle
- Dominique
- Easter Egger
- Cochin
- Buff Orpington and Delaware

HOT CLIMATES

- Silkie
- Welsummer
- Leghorn
- Easter Egger
- Egyptian Fayoumi
- Blue Hamburg
- Black Sumatra
- Barred Rock Bantam
- Barred Plymouth Rock

What Nourishment Does Your Flock Need?

Every animal has nutritional needs that you should meet. Nutrition is considered the process where animals take everything in, food, and use it for health and growth. You must meet your nutritional needs for your flock to function and serve you properly. You should ensure you have the right chemical compounds in the food and serve your flock with the right amounts. An animal can't produce all the components required to ensure health. Also, understand that you cannot continuously feed your flock with one type of food. No food is perfect and contains all the components. Therefore, you should feed your chickens with all the supplies.

CARBOHYDRATES

Dietary carbs come alone, in pairs or large forms like plant starch. An example of alone carbs is glucose and in pairs is sucrose. Like humans, chickens use carbs as a source of energy and fuel source in cells. Carbohydrates are usually the largest component in chicken feed. These carbs mostly come from wheat, corn, sorghum and millet. These carbs come in two forms: digestible (like starch) and non-digestible carbs (like cellulose). Indigestible carbs have important roles in the chicken body. For example, it helps with intestinal health but be careful not to overdo it as it could interfere with growth and give the chickens intestinal problems.

PROTEINS AND AMINO ACIDS

If you want to identify a chicken feed with proteins, check the label, it's marked as "crude protein" A protein is a combination of about 20 different amino acids. Your bird requires ten of these amino acids as it can produce the other ten on its own, ensure your feed has these amino acids. We must let you know that it's difficult to get all the amino acids in one type of food, such as corn or soya

beans, but a mixture of the two approaches is a good balance. Some of the benefits of protein in the chicken body include:

- Enhance growth
- Egg production
- Immunity
- Adaptation to the environment

VITAMINS

Vitamins are classified into two categories, fat-soluble, which include vitamins A, D, E and K. The water-soluble vitamins include vitamin C and B-complex. Vitamin A in the feed gives a golden yellow York, which consumers want

for their eggs. Vitamin C is not necessary for the food, but if your flock is under stress, adding it to the diet is a good idea. Understand your flock's vitamin needs and feed them accordingly. They will reward you.

WATER

Water is necessary for any living thing, but it is often overlooked. Your flock can live longer without food than it can without water. The amount of water your chickens need to live a happy and healthy life depends on the temperature of the place they are in, growth rate and egg production levels. Let's not forget your chicken's ability to reabsorb water through its kidney. Reduced water intake, even for a few hours, could reduce the level of egg production; therefore, it's advisable to always have clean water for your flock. It is recommended to have a waterer to ensure your chickens never run out of water. If you are not using automatic waterers, fill the drinkers twice or thrice a day, depending on the size.

FATS

Unlike carbohydrates which provide four calories per gram, fats provide nine. Fats are two and a quarter times heavier in calories than carbohydrates. There are two types of fats: saturated fats, solid at room temperature, and unsaturated fats, liquid at room temperature. These are some saturated fats that you can use with your poultry tallow, poultry fat and lard. For unsaturated fats, use corn oil, canola oil and soil oil. For your chickens to absorb fat-soluble vitamins A, D, E, and K, they require fat in their diet. You should know that fats in feeds can go bad and become rancid, and the chances are higher in summer. To prevent this, add antioxidants like ethoxyquin.

MINERALS

Minerals are classified into two broad categories known as macro and micronutrients. Your chickens need more macronutrients and fewer micronutrients. Although poultry requires lower levels of micronutrients, the nutrients play a vital role in body metabolism. Let's take iodine, for example. It produces thyroid hormone that regulates energy metabolism. Micronutrients include iodine, iron, copper, zinc, selenium and manganese. The macronutrients include phosphorus, potassium, calcium, chlorine, magnesium and sodium. Calcium is an important nutrient that helps bone formation and eggshell quality. It also helps in muscle contraction and the formation of blood clots. The last two functions are less known but very crucial in the body. Phosphorus aids in bone development. Chlorine forms hydrochloric acid in the stomach that helps digestion, sodium and potassium are important in muscle, nerve and metabolic functions, and magnesium helps in muscle and metabolic function.

Is Chicken Raising Legal?

Chicken keeping laws vary from one area to another. A certain state will allow it, while another one will not permit it. Zoning regulations combine with the laws each area. Although this is the case, these laws can change anytime. Since the question of whether chicken keeping is legal depends on your area, there are key issues you need to understand. Sometimes the local authority could permit a simple number such as four or more. Sometimes the authorities permit chicken keeping in rural or agriculturally-designated zones but not in suburban or urban zones. Some areas may ban roosters but permit hen keeping. Your chicken coop requires you to pass the local building regulations. Permits are payable annually. In some areas, it is illegal to slaughter a chicken. Almost all urban areas have laws relating to noise. Coops should be far from the roads and nearby residents in the cities. Before purchasing your flock, check the zoning laws, lease restrictions, building codes and any restrictive covenants.

How Many Chickens Can You Keep, and What Are the Factors to Consider?

The number of chickens you keep will depend on the purpose or reason you are raising. If it is a hobby, you do not require a big number compared to raising chicken for business. If it's pets, then you need about three chickens. If you want eggs, then ten or so will be enough. If you want to sell the eggs, the number will go up to 15 chickens. You could also determine whether you need a rooster or not. If your aim is selling eggs and meat, then a roaster will be necessary. For pet keeping, you could go for a breed that lasts longer. Some go for about ten years and some for three years. There isn't the right number. The number of chickens depends on why you need chicken.

FACTORS TO CONSIDER?

- Start with a small flock if you haven't raised chickens before.
- The size of your backyard.
- Local authorities, what does the local authority do about raising chicken in your area?
- Personal egg consumption, if you have a large family and eat a lot of eggs, you should keep a bigger number of chickens.
- Your coop size.
- Your budget.
- The breed and size of chickens you want to keep.

WHAT'S YOUR BUDGET?

You may think that the chicken raising project is expensive. That can be true, but only when you are starting it. As a beginner, you will require to purchase many things such as a coop, waterer, brooder and other equipment. Food is also required. Although you will purchase all these, it doesn't mean the cost will break your bank. You will manage. After this big step, the expenses are less, and you will enjoy having them around. If you raise the chickens as pets, they will be low-cost pets that feed you with eggs. Let's estimate what it might cost you for a start-up. The number is four chickens. Remember, the price may differ from one country to another.

Item	Price
Coop	$200-$2500
Feeders and waterers	$20-$70
Nesting boxes	$37-$300
Chicken (4)	$8-$320
Fencing	$30-$700
Chicken feed	$12-$36 per month
Chicken bedding	$5-$50
Perches	$20-$40
Miscellaneous	$0-$500
Total costs	$332-$4516
Item	**Price**
Chicken feed	$12-$36 per month
Chicken bedding	$5-$50 bale
Miscellaneous	$5-$50
Total cost	$22 minimum (up to $100)

The expense is higher when starting the project and reduces after that. Start-up expenses will also differ with the type you want to raise, the number of chickens, and the size of the coop to be bought. As you can see, the cost is not that, and you shouldn't refuse to start thinking that it will always be expensive. The benefits of keeping them are way much better.

Mash Feed

Crumble Feed

Pellet Feed

Chicken feeding simply means giving chicken food. Once you've brought your chicken home, the feeding does not stop. Proper chicken feeding reflects on the chicken products. If you feed your flock properly, its eggs and meat will show that it is or was a properly fed bird. The ratio is also important during feeding. In case you feel that you are not sure how that is done, you can talk to an expert in this field. The best way to feed your flock is with a balanced pelleted ratio. It doesn't matter the kind of lifestyle, whether confined or free-range. Almost all of these diets are balanced. They have soybean for protein and corn for carbs to give energy, vitamins and minerals. Chicken commercial feeds contain antibiotics and arsenic, which promote health and improve growth. It also sometimes contains mold inhibitors. If you do not want any feed containing these inhibitors, you can check the labels for one without. Chicken feed is pelleted to make it easier for the chickens to feed. They eat more at a time. Chickens are nibblers; you will see them make several trips to the feeding trough to eat. Pelleting makes feeding easier. The chicken eats more at a time, reducing the energy required to feed. It is crucial to note that different stages of the chickens' growth require different ratios of nutrients. For example, a younger bird requires food with a higher ratio of proteins while bigger chickens require lesser. If you do not prefer commercial feeds, you could do a home mix in the comfort of your home. This is a good choice to ensure a natural mix with no additives. Some of the ingredients you can use include the following, for energy-oats, corn, sorghum and barley. Proteins-soybean meal, sunflower, peanut, sesame, safflower, animal products such as meat, fish, and grain legumes. It requires a great effort to ensure that a home mix is properly balanced, especially the certified organic diets.

Forms of Chicken Feed

- **Crumble** is a coarse variety of mash that is not as compact as pellets. They are semi-loose and easier to manage than a mash. Crumble can bridge the big gap between the compact pellet and mash.
- **Pellets** are the most common chicken feed. They are compact and cylindrical. They are usually the first choice amongst chicken keepers as they are easy to dispense and reduce food wastage.
- **Mash** is an unprocessed and loose chicken feed. Mash feed has a very smooth texture and is the finest chicken feed. It is mostly used for baby chickens. Although it's a chick's feed, your adult chickens can still safely eat it. Mixed with water, it creates a great treat for your chicken. A huge disadvantage of this form of feed is that it's prone to incidental waste.

Types of Chicken Feed

STARTER CHICKEN FEED

This food is prepared to meet the nutritional requirements of chicks. It is high in protein, about 24 percent. Generally, chicks are given starter feed and water until they are six weeks old and ready to progress to grower feed. It's not advisable to continue feeding your chicks with the same amount of the starter after six weeks, as it could lead to liver damage.

GROWER CHICKEN FEED

In simple words, this is a feed for teenage chickens around six to twenty weeks old. The dietary requirements are different from that of a baby chicken. The protein content goes down from twenty-four percent to about sixteen to eighteen percent and has less calcium than the regular layer chicken feed. This food enables your chickens to grow without overburdening them with unnecessary minerals and vitamins.

LAYER CHICKEN FEED

This feed balances all the nutrients necessary for egg laying: proteins, calcium, minerals, and vitamins. Proteins levels are the same as those in a grower feed, but extra calcium is added to help with the eggshell formation and ensure that shells are crispy, crunchy, and clean. Feeding a baby chick or growers this type of feed will not meet their needs; therefore, a layer of feed should only be given once a chicken has started laying eggs or is twenty weeks and above.

MEDICATED AND UNMEDICATED

Medicated chicken feed contains chemicals to treat coccidiosis and other flock diseases. Unmedicated feeds lack these additional chemicals. It is mostly fed to starters and growers. A medicated chicken feed contains a chemical known as amprolium, which helps protect your flock from deadly diseases. Remember not to use any medicated feeds if your flock is vaccinated.

FERMENTED FEED

Fermenting chicken feed helps improve vitamin and enzyme contents in the food. It also makes the food easier for your flock to digest. Due to its density, the chickens remain fuller for longer. Fermenting your chicken feed will reduce feed expenses and chicken droppings.

BROILER FEED

This feed is for chickens being raised for consumption. The feed contains a higher ratio of protein that helps the chicken grow faster and bigger. Note that you shouldn't feed layers broiler feeds as the protein is too much and not always beneficial. Broiler feed varieties consist of a starter, grower, and finisher.

CHICKEN SCRATCH

Chicken scratch is not a type of chicken feed you can feed your flock daily, and it should rather be viewed as a treat. Although scratch isn't as nutritious as other feeds, its abundance of cracked corn and other types of grains is sure to make your flock happy. Additionally, chicken scratch is a good energy source and helps keep your chickens warm during cold seasons.

How Do You Store Chicken Feed?

Store your chickens' food in a cold, dry place and away from wildlife. Chicken feed is packed and sold in sacks, which means they are safe at the time, but when you open it to feed your flock, the remaining feed becomes exposed to other elements and can be damaged by humidity. So, how do you ensure that the 50 pounds sacks you bought won't go bad? The answer is proper storage. Ensure when you are doing so, you still maintain its nutritional value. Follow this simple guideline.

- Be careful with the amount of feed you store. Chicken feed has a shelf life which is usually a few months. The amount of food you purchase and store will depend on the size of your flock. Although it is tempting to buy a lot of feed to avoid many shopping trips, you risk your stored feed going bad before it is eaten. Put old and new feed in different containers to avoid cross-contamination if the older feed goes bad.
- Pick the right type of storage container. Apart from having the right size, ensure the container is airtight and made of a material that protects the feed from rats and mice. Metal and heavy plastics are a great choice. Although the plastic can be compromised, it will take a while before that happens, around six months. Another consideration when using plastics is to make sure they are opaque to protect the vitamins in the food from direct sunlight.
- Location is key. Store the feed indoors and in a ventilated room away from moisture, sunlight, and wildlife. Keeping it off the ground can prevent condensation problems. If you prefer storing your food outside, find a shaded and dry location away from harmful elements.
- Ensure it's clean. Your storage room should always be clean. Sweep it regularly and clean empty storage containers before adding new food.

Here are some of the things that can ruin chicken feed:

- Moisture
- Light and heat
- Chemical reactions
- Time
- Pests

WHAT TO AVOID

Although chickens are not picky, there are some foods that should be avoided. Anything high in fat or salt is not advisable. Rancid or spoilt food should never be fed to your chickens. Other foods that shouldn't be given to your chickens are:

CHOCOLATE OR CANDY

Theobromine is a compound found in cocoa- an ingredient in chocolate and many candies. It is well known that this

ingredient can be lethal in dogs and cats, and the same can be said for your flock. A piece of cholate could cause heart problems such as irregular heartbeat or a full heart attack. The lethal effects of chocolate can be realized within twenty-four hours. The darker the chocolate, the more of this compound it has and the more dangerous it is.

AVOCADOES

The pit and peel of avocado have persin which is toxic to chicken. Although many have fed their flocks avocado with no complaints, it is not recommended, but it's not advisable due to the potential risks. Too much persin will cause heart problems and difficulties breathing, leading to eventual death within forty-eight hours.

DRY BEANS

Unlike cooked beans, which are safe to eat, dried beans are lethal. Dried beans contain hemagglutinin which is very dangerous to your chicken. They can cause serious illness to humans, and your flock is not an exception. Although all dried beans are fatal, kidney beans are the most potent, with illness from as little as three beans. Unfortunately, little can be done following the consumption of these dried beans.

MOLDY FOOD

You wouldn't want to eat moldy food, and neither should your chicken. The mold that grows on soft fruits contains toxins. Aflatoxin increases the chances of developing liver cancer in humans and animals.

GREEN POTATOES

Feeding your flock cooked potatoes is okay if done as an occasional treat. However, avoid green potatoes, especially in large quantities. Green potatoes contain a toxin known as solanine which is poisonous to your flock. Cooking methods, which include boiling, do not do anything to remove or alter solanine, and therefore, it is best to simply throw away your green potatoes.

JUNK FOOD

Junks are not good for your health or your flock's.

HIGH SALT FOODS

Foods with high salt content will cause eggshell deformities over time. Salts help in growth and development but only in moderation.

SUPPLEMENTS AND HERBS

Herbs are inexpensive to grow and provide different health benefits; they act as antitoxins, natural dewormers, and antibiotics. Some herbs help with respiratory function and boost the chicken's immune system. It's good to start feeding your chicken with herbs before they fall sick. Waiting until you see the symptoms could be dangerous, as some chickens are good at masking their symptoms. In short, prevention is better than cure. You can use herbs in all areas of chicken raising, that is, in the feed, brooder, water, nesting boxes, coop floor, and dust bath. Here are some of the herbs that you can use in your journey.

- Lavender acts as a stress reliever and increases blood circulation.
- Marjoram helps with blood circulation, egg laying, detoxification, and inflammation.
- Sage acts as an antioxidant, antiparasitic, and health promoter and combats Salmonella.
- Basil is an antibacterial and helps with mucous membrane health.
- Marigolds will help your chicken with blood circulation, parasite treatments, and relieving stress.
- Garlic assists with blood circulation, egg production, and preventing fungi infections.
- Parsley acts as a laying stimulant and helps with blood vessel development.
- Mint is a rodent repellent, insecticide, and antioxidant. It also helps with digestion and lowers body temperature.
- Nasturtium can act as an antibiotic, insecticide, and laying stimulant.

SUPPLEMENTS

Supplements boost the immune system, especially during winter, summer, and molting.

- Grit is small pieces of gravel that help break food into small pieces; it is suitable for confined chickens which do not go out to pick dirt or gravel.
- Calcium helps with eggshell formation.
- Vinegar helps with respiratory problems and digestion.
- Electrolytes act as stress relievers.
- Probiotics boost the immune system and digestion.
- Molasses is a source of iron and minerals.

WATER MANAGEMENT

When it comes to water, quality and quantity are important. Water should be clear, odorless, and tasteless. Chickens can go longer without food than they can without water. Insufficient water will cause early molting, growth delays, stress, and alter the egg-laying pattern.

Automatic watering units are a great option if you can afford them. If you're looking for something less costly, water fountains serve as a source of drinking water and can cool your chickens down during hotter months.

How do you ensure proper water management?

- Have an adequate water supply.
- Invest in efficient water houses.
- Have the right cooling system.
- Use efficient drinkers.
- Manage water quality and temperature.

Identify contaminated water

- Water becomes cloudy.
- Water may turn reddish-brown.
- A blue hue to the water.
- The water smells like a rotten egg.
- The taste of the water changes.

Tips for water management

- Test the water semi-annually.
- Inspect the water from time to time. A frequency of once a month is recommended.
- After testing, make sure a vet interprets the results.
- If you source your water from a borehole, ensure it's safe for your chickens by testing for trace elements such as manganese and iron.

Routine management (Daily, weekly, and monthly activities) Your chicken's health, safety, and nutrition matter greatly; therefore, you must have a routine to ensure everything runs smoothly. Your flock needs your attention and care, just like little children. You shouldn't make it too complicated as you may eventually get tired and give up on the project, as it may seem a challenging task for you.

DAILY ACTIVITIES

These are the most important activities for a happy and stress-free keeper; you shouldn't skip them even for a single day if you want a happy and healthy flock.

- Water check. Water is an essential element of life. Your flock needs to drink a lot of water daily as they are thirsty animals and easily get dehydrated. The chicken will not drink contaminated water; therefore, you should change it daily to ensure your flock is happy. Chicken water becomes easily contaminated by droppings, bedding, and many other contaminants in the coop, so it is necessary to make checking your water a daily activity. When changing the water, clean the drinker to remove any slime or algae that may have formed in the container. A more expensive option is an automated drinker, which prevents contamination as it is computerized and allows chickens to drink from the drinker's nipples instead of from an open container.

- If you want a happy flock that welcomes you home every evening, you need to meet their daily food requirements. If you eat every day, make sure your flock eats too. Ensure your chickens are getting adequate amounts of clean and healthy food. If it's a starter or grower chicken, give it food that meets its nutritional needs in the right quantities.
- Egg collection should be a daily routine. Do it in the morning before they are broken or contaminated by your chickens' excrement. Dirty eggs are difficult to clean and could end up breaking.

- Flock watch. Chickens love attention, especially if you are keeping them as pets. If you don't check on them, they will feel neglected. Spending time with them will form a strong bond, and you will learn how your flock behaves and notice when something isn't right. Take time to notice any kind of bullying, frustration, illness, or injury. Attentive keepers are good keepers, and your chickens will run up to you with love when they spot you.

WEEKLY ACTIVITIES

- Coop cleaning and maintenance. You will notice that by the end of the week, the coop becomes dirty with poop and rotting food. You should clean the coop weekly. Cleaning can be a fun activity as it is an opportunity to bond with your flock. You will see the chickens scratching the floor as if helping you clean. Let them do it; they are enjoying your company. A clean coop makes your flock happy.

MONTHLY ACTIVITIES

- Bedding cleaning: How often you clean your bedding depends on the type you are using, but make sure that you take time to do it occasionally. Remove rotten food and any other dirt that could compromise your chickens' comfort.
- Nesting box clean-up: You may remove any noticeable waste when you collect eggs daily but do a thorough check monthly to clear anything that you hadn't noticed before.
- Drinker sanitization: Chickens need clean water. Check for any contamination and blockages and fix them.
- Weather changes preparation: Check what your chickens need to protect them against harsh weather.
- Coop sanitization: This is challenging, but it simply has to be done. You need to move the flock somewhere, which they might not like, clean all the areas that need cleaning and use a safe sanitizer, such as diatomaceous earth, which can be sprinkled on the floor.

QUICK CHAPTER TAKEAWAY

In this chapter, you learn about the different feeding and storage methods, what foods you should avoid feeding your chicken, how to manage their water supply, and routine management.

You should understand coops and the different varieties in the market in order to choose the right one for your flock. A coop is necessary to ensure that your flock is healthy and happy. It gives them a place to sleep at night and roam when the sun rises. A flock without a coop will find its own place to sleep and lay eggs, which could be open to predators.

What Are the Benefits of a Coop?

- Protects chickens against predators. Although chickens are very territorial and willing to fight when attacked, there are some predators that they will inevitably lose to. Coops allow your chickens to relax peacefully and not always be on the lookout, wondering what might happen to them.
- Protection against harsh weather and dangers during the night. Chickens can easily adapt to environmental conditions, but this doesn't mean that they should be left without a home. A coop will ensure that they will be safe inside when night comes and the chickens are mostly blind.
- Chicken supervision. A coop will make it easier to watch your flock. After a day of fun and roaming around, your chickens come back at night, and it will be easy for you to do a head count and see if there is a missing or injured chicken.
- Ease of gathering eggs. If chickens don't have a coop, they will lay eggs anywhere they feel safe. This creates a difficult situation for the person gathering eggs. Coops give your chickens a safe nesting place, and you will no longer have to play hide and seek with the eggs.

Types of Chicken Coops

WOODEN CHICKEN COOPS

Wood is a cost-effective material. You could get a fabricated coop

and assemble it yourself, or you could build it from scratch. Wooden coops come in different shapes and sizes. The following are some of the wooden coops available in the market.

- The Combined coop is a coop with an extended run. This coop enables your flock to roam in a protected area (the run). If the coop has a wheel, it will be possible to move the chickens around the compound to have a different view daily.
- The A-Frame is an easy-to-clean coop and a good option to teach your children chicken farming. It enables your children to care for the flock.
- The Quaker has an attractive appearance and protects the flock from predators.
- The Dutch coop provides security and beauty. It looks attractive in the middle of your yard and is a good choice for small flock keepers.
- Like any other coop, the Lean-To coop protects chickens from predators and makes it easier to gather eggs.
- The tractor chicken coops are easy to move around. This movement helps reduce the number of bugs in your backyard.

PLASTIC CHICKEN COOPS

These coops are made from heavy-duty plastics and hence are very durable. They are portable and easy to clean. Some disadvantages of this coop are that they have very small runs that do not give the chickens enough space to move around, they have to be placed on flat ground to ensure that there are no weak points that could let in predators, and they have poor ventilation. These coops are cost-effective, easy to clean, portable, and easy to assemble. When purchasing a plastic coop, these are some of the factors that you should consider. The size of the coop, ease of set up, ease of cleaning, weight (a light weight is better if you will be moving it around on your own), roosting bars, material, security, and ventilation. The following are some types of plastic coops.

- Snap Lock is the most common brand for large plastic chicken coops. Their coops are durable and easy to set up, although you must build your stand.
- Omlet Eglu Cube Large Chicken Coops are the best plastic coops but can only accommodate a small number of chickens. It takes about seven hours to assemble the coop, although some say it takes more.

ECO CHICKEN COOPS

An Eco coop is made of recycled materials. It is an Eco-friendly coop with optimal durability and is easy to clean. It does not come with a run, so you can build one for yourself and determine the size you want. It's a good choice to purchase as it's easy to ship or transport. The following are some Eco chicken coops.

- Solway Deluxe Mini Hen Coop is ideal for two to five hens. It is moveable to suit your and your chickens' needs. It has two nesting boxes and a chicken run.
- Solway Eco Hen House is suitable for three to six chickens.
- Solway Deluxe Eco Hen House has the option of having front feeder holes or a large opening door, making feeding easier.
- Solway Eco Hen Ark is suitable for keeping more chickens.
- Solway Deluxe Eco Hen Ark is suitable for a maximum of ten chickens.

CHICKEN ARK/CHICKEN TRACTOR

A chicken ark is a small coop designed to keep the chickens enclosed and give them room to move around. These coops have no floors and are moveable. They are suitable in areas where the local authorities or predators do not permit chickens to move around freely. It is also preferred for small-scale farming. Less than twelve chickens can fit in an ark.

- Barn tractors are bigger than the other coops. It is an elevated coop with an extended run. These coops offer a comfortable home for your flock as it has windows and a door that helps with proper ventilation.
- A-Frame chicken tractor is easy to make and a functional coop. It is suitable for layers, and holes in its roof give good ventilation. This coop can use a lock system to ensure your flock remains safe.
- Chicken tractors with roost bars
- Hoop-style chicken tractors have unique designs. They are dome-shaped, unlike the common triangle or box designs. Although it may look complex, it's an easy model to build. These are made of a lightweight material that makes them easy to move around.
- Chicken tractors with bottom floors

- Box-style chicken tractors
- Chickasaw tractors
- Geodesic chicken tractors have wheels and are easily moved. They are also made of strong and light material.

Traditional Chicken House

This coop, designed by Tarter Farm and Ranch Equipment, dates back to 1895. The coop has a large wooden structure, indoor roosting space, and a door leading to a fenced-in run. The houses are categorized into three systems: Free-range or extensive, semi-intensive, or intensive. Here are some types of traditional chicken houses.

- Brooder house
- Grower house
- Layer house
- Broiler house

POULTRY SHED

A poultry shed has an easy design and is easy to build. It is enclosed and has several nesting boxes, roosts, and feeders. It also has a sloping room and front windows, which allow natural sunlight in the coop. Poultry sheds allow outdoor access to your flock by building a fenced-in area around the coop.

PORTABLE COOPS

This coop does well with modified free-range chickens as you can move the coop easily around alongside its run. The portable coop provides safety for your chickens. They can roam around as much as they want and still have a place to return to at night. These coops can be made of different materials, and mostly they are built on skids or wheels, making movements easier.

Custom and Homemade Coops

A custom coop is a chicken house that is purchased and ready for you to assemble, while in a homemade coop, you build the chicken house yourself from scratch. A readymade coop is easy to assemble but does not give you the room to build your coop how you like it.

WHY CUSTOM COOPS?

- Less time-consuming. Building a coop can take a lot of time. You have to start by researching the available designs, decide which one meets your needs, and then draw a plan. As part of the research, you must look for the required materials and ensure they are up to standard. After building the house, other things need to be built too, for example, nesting boxes, the run, and perches. If this sounds like a stressful amount of time and energy, you might prefer a readymade one.
- An easy and relaxed time. The world we live in today does not allow us to have too much free time, especially since most people work from Monday through Friday. People have duties that require their attention, and even during weekends, they do not get time to relax. Although your weekends could be free, would you prefer to relax and prepare for the coming week or start building a chicken coop from scratch? Some will find building a fun and engaging activity, but they are also those that want an easy time and would prefer to purchase one, and if not, they could give up on the project altogether.
- Less expensive. Building a coop is expensive, and you might not know how much it will cost you unless you have done it before or know someone. When you walk to a shop to purchase a coop, you can always compare prices as there are several options to pick from, you can choose one that suits your budget, and that will be a done deal. With building, you will have to purchase materials like wood or any tools that you don't already have, such as an electric gun.
- Equally comfortable. A commercial chicken coop has all the requirements for your flock to be happy. They have nesting boxes, perches, a run, and a good ventilation system. There are different designs and sizes that you can choose from. Whether you buy or build the coop, the flock will be happy in their home.
- Determine whether it's of good quality or not. The popularity of commercial coops means that there are many reviews out there. Additionally, if this is your first flock, buying a coop to start with will help you understand how you could build a better one for you and your chickens' needs.
- Coop features will perform properly. Commercial coops are built in a way that meets certain standards; they are also tested before being sold to the consumers, unlike homemade ones where you are not sure if they will work as expected.

WHY HOMEMADE COOP?

- You can customize your coop. The major benefit of building a coop is building it how you want it. You use your imagination and expertise to build something unique without following instructions.
- It can be a source of income. You can use your skills to make an income. Most times, people don't want to spend hours and days building a coop. If building your own coop is a good experience, you can branch out and build coops to sell for additional income.
- Building your coop can be cheaper. Buying a coop is convenient, but building can be cheaper, especially if you have the idea and skills to do it. You could go to a good supplier, get the materials at an affordable price, and build it at no added cost.
- Family bonding time. Building takes time, and it could be a good opportunity to spend time with your family catching up and sharing ideas.

HOW TO BUILD YOUR CHICKEN COOP AT HOME

Building a chicken coop is an easy and fun activity. For some, it is a hobby. The truth is that it can be challenging without the skills or a guide to help. We have simplified it for you by providing a simple guide to follow.

- Build frame ends, pieces, and floor. To start building, tackle the floor first, put up the joists to a single stringer, and shoot nails through to the end of each joist. The frame will be complete.
- Add plywood floor, trim the end of the floor, and use a reciprocating saw to cut the notches in the corner to fit around the posts. Put the plywood on the frame and staple the plywood.
- Add supports and add horizontal and vertical supports to the sides. Attach the sides using brackets and pocket holes. Create the pocket holes using Kreg Jig or freehand pocket holes with a drill.
- Add support (continued). Next, you should add the verticals on the front end. It will support the nesting box.
- Frame upper walls. Frame the upper walls using horizontal and vertical measurements. Use whatever you have. Any size will work. Use screws to attach the upper frame to the lower frame.
- Frame upper walls (continued). All angles are 30 degrees. Ensure the roof supports the overhang on the bottom. Attach the roof to supports and then to the posts and the verticals with grabber screws.
- Frame upper walls (continued). The roof trusses are on each end and in the middle. The roof pitch is usually 30 degrees. Cut one end of the boards at 30 degrees and leave the other end square. The middle is cut slightly differently, with 30 degrees on the end and 60 degrees on the other. This end and the same piece for the upper roof are the only ones that are not 30 degrees.
- Framing hinged sides, now the roof is ready. There is still one more piece until the frame is complete. This is

where the hinge for the sides attaches. Once the piece is in place, the framing is complete and starting to look like a coop.

- Attach siding. When everything is already framed, add the siding. Cut the plywood to size before using a brad nailer to attach it to the coop. At this stage, only attach the front, back, and one side. Attach a side that you will open up later. All angles on the roof are 30 degrees.
- Create siding for the side that hinges down, put together two pieces of the plywood siding, and then frame the outside. The completed dimensions will be slightly smaller than the other side to allow the side to hinge. Set the door aside until it is time to paint. If you have time, you might want to paint the siding before you attach the trim to the outside.
- Attach plywood roof. By this stage, the coop is almost complete.
- When building a nesting box, the box measurement will depend on the coop size and the box that can perfectly suit the coop. All angles in a nesting box are 20 degrees.
- Building nesting box (continued). Cut the plywood siding and attach it to the frame. The sides of the box will be notched out. Use the notch to slide the nesting box into the hole in the coop and attach the plywood sides to the inside you already have in place on the front wall of the coop.
- Paint the coop. Add your preferred color to make it look attractive. Paint the trim boards before you start trimming.
- Attach the nesting box. When you finish painting, add the box. Cut a hole in the front for the nesting box to attach to the coop. Cut the plywood siding using a reciprocating saw. Ensure you double-check the measurements.
-
- Attach the nesting box (continued). Place the nesting box on the floor and frame of the chicken coop. Cut the nesting box from the bottom. When the nesting box is in place, screw the box to the coop using a grabber screw. The view from inside will show how the box attaches to the coop. Ensure the nesting box is sturdy.
- Add a roof to the nesting box. The nesting box should hinge open. Screw the piano hinge onto the edge of the nesting box roof first, and then attach it to the nesting box.
- Add a fold-down side. Once the box is complete, add the side that folds down.
- Trim the chicken coop. Make sure it is painted before trimming and that the measurements are correct.
- Add lamp. Cut a hole in the back from the bottom and insert the lamp. Make sure it is properly supported.
- Add roofing material. Metal roofing is less expensive than shingles, easy to install, and durable. Build your chicken coop using measurements that suit your needs.

Factors for Consideration when Choosing a Chicken Coop

THE SIZE OF THE CHICKEN COOP

The size of the coop is determined by the number of flocks a farmer wants to keep. The larger the number, the bigger the coop. Size is a basic consideration when purchasing or building a coop. A standard coop has at least four square feet of space for each chicken and ten feet squared for the run. Adequate space in the coop will prevent the spread of diseases from one chicken to another and enable socialization. Overcrowding Is a health hazard for the flock, especially if one has an infectious disease. In addition to that, overcrowding can make feeding and sleeping difficult. The chickens can trample each other when reaching for food, increasing the chances of getting hurt.

BASIC CHARACTERISTICS OF A COOP

The wide variety of coops out there all have different features, but all good coops should meet some basic requirements to make your chickens happy and your life easier.

AVAILABLE NESTING BOXES

If chickens don't have a nesting box for laying, they will lay their eggs anywhere. Before purchasing or when building a coop, ensure it has nesting boxes to make egg laying and collection easier.

ELEVATED THE CHICKEN COOP

A proper chicken coop should be elevated to ensure that predators cannot dig underneath to reach the chicken in the coop. Elevation also prevents other animals, such as rats, from building homes in the chicken coop and supports proper air circulation, giving your flock some fresh air.

MEETS YOUR BUDGET

There are different coop designs and sizes, be prepared, and don't over-estimate your budget. Coops are expensive, but you can always find one that fits your budget.

COOP QUALITY

No one wants to spend money to buy a coop that will fall apart after a while. If you notice some gaps between junctions, the coop is not warm enough, and your chickens will have a hard time during winter. If the coop shakes when you push on it, it is not steady enough to meet your needs. Check if there are any rusted parts to avoid the expense of replacing them.

MOBILITY OF THE COOP

There are stationary and mobile coops. A stationary coop will work well for you if you own a home. If you are not a homeowner, a mobile coop might be better. For added mobility, you should consider a lightweight material that will be easy to move.

THE CLIMATE

Your area's climate will determine the type of coop to buy and how and where you put it.

HAS A RUN

When buying a coop, make sure you get one with a run. If you already have one that doesn't have a run, don't worry. This can be fixed, although it involves added cost. Buy a chicken wire and wood to build a run for your flock.

EASE OF CLEANING

Cleaning a coop shouldn't be a challenging task. Removable parts should be easily removed and returned without causing injuries or breakage.

AVAILABLE CHICKEN FEEDERS AND DRINKERS

Feeding is a daily activity, and these should be available in the coop to ensure your flock is happy.

AVAILABLE STORAGE SPACE

There should be an area designated for food storage. The storage should be dry to prevent food from going bad. If the food rots or grows mold, it will be a health hazard for your flock. The storage should also be safe from other wild animals.

PROVIDES ADEQUATE PROTECTION

The coop should have proper roofing and be elevated to protect against air and ground predators.

Tips for Cleaning and Maintaining Your Chicken Coop

Cleaning your chicken coop is generally a weekly activity, although this may depend on the size of your flock, the weather, and the amount of waste they create. It is wise to remove any chicken droppings daily, especially in the mornings when you go to collect eggs. This will ensure that by the end of the week, the coop will not be extremely dirty and will be easier to clean. Cleanliness has a very important role in chicken raising. It ensures that the flock remains healthy and comfortable, especially the baby chickens. A daily clean will not be as thorough as a weekly clean. Here you should clear all the chicken droppings, change the bedding, and use white vinegar to clean and ventilate the coop. To easily clean the coop, the door should be large enough for you to go inside or removable. It's also helpful to be able to remove the roof, nesting boxes, and perches during cleaning. If you cannot remove the nesting boxes and the perches, use a hose to clean them.

- Have a weekly schedule. Having a cleaning and maintenance plan makes it harder to neglect your flock. Having a plan will help you be prepared and remember to perform cleaning activities.
- Fix the broken area. Mending broken areas is crucial as it prevents any injuries to your chickens or attacks by predators. It also ensures that the flock has a constant water supply and food. If the water supply is broken or blocked, it will risk your flock's health.
- Evacuate the flock. Chickens are prone to cause distractions

- Disinfect the coop and nesting boxes. Use natural agents such as vinegar, and avoid using any bleach as it can be toxic to your flock. You can create a solution by combining equal portions of water and vinegar and use it to clean the coop. You can also clean drinkers and feeders using the vinegar solution, but make sure you leave them outside under direct sunlight to dry completely.
- Give the coop some time to air dry. Direct sunlight will act as a disinfectant. You can let the coop dry as you dry the nesting boxes.
- Move everything back inside the coop. Make sure you change the bedding, assemble everything you had removed, and put the feeders and waterers back. Give your flock a check and put them back one by one. Please make sure they are healthy and happy. Make sure you wash your hands after finishing everything.

ELEVATION OF A CHICKEN COOP

You can place your chicken coop directly on the ground, but it is advisable to have a base. You can provide a small elevation by using crushed stone as the base for a wooden coop. Although this will work, a minimum elevation of three feet is recommended. If you don't elevate your chicken coop, your flock could be exposed to drafts during roosting and be too cold during the cooler seasons. When searching for a location for your coop, pick a location with direct sunlight. Natural sunlight is very important for raising your chicken. There are some exceptions that will prevent you from elevating your coop like if your coop doesn't have a floor or the coop has a concrete floor.

and slow you down. Remove them in the coop and anything else that could get in the way of your cleaning.
- Clean during a warm day. You do not want to leave your chicken outside during a cold day to clean the coop. This is risking your health and that of your flock.
- Check on your chickens. This is a good chance for you to inspect your flock, give each of them a look and see if there are any injuries or signs of frustration. It is also a good time for you to cuddle your flock, give them your attention and bond with them.
- Sprinkle some fairy dust. Sanitizing and disinfecting the coop is necessary. Something like wood ash and lime is great at reducing coop odor.
- Put on your mask when cleaning. The coop is contaminated with chicken skin cells carrying campylobacter, other bacteria, and fungi. You should wear a mask or respirator to protect yourself.
- Use a misting bottle to dampen the floor and prevent dust particles from flying.

A STEP-BY-STEP GUIDE TO CLEANING A CHICKEN COOP

- Take the flock out of the coop and put them somewhere safe for some time.
- Remove everything from the coop so that you don't have to clean around them, meaning all drinkers, perches, and nesting boxes. Once this is done, scrape out the chicken dirt, cobwebs, and any unnecessary thing from the coop. Make sure you have your mask and gloves on.
- Using your hose, spray the enclosure down to remove the dust created by the previous step.

WHY SHOULD A CHICKEN COOP BE TALLER?

- Roosting bars should be high enough for the chickens to walk underneath.
- Roosting bars should be higher than nesting boxes. If the boxes are higher than the bars, your flock will sleep in the boxes as chickens like roosting on the highest spot in the coop.
- The ventilation system should be high enough to prevent drafts. You shouldn't expose your chickens to drafts, especially on roosting bars at night.
- Coops need to be elevated for adequate ventilation. If your chickens are in a humid area, you need more ventilation. A large flock requires more ventilation too. A coop with a high ceiling requires less ventilation than a low ceiling.

WHY WOULD YOU NEED A TALLER CHICKEN COOP?

- When you have a large flock.
- Your chicken run is too small.
- Your flock consists of large chicken breeds or roosters.
- You are using thick bedding for your chickens.

Benefits of Chicken Coop Elevation

PROTECTION FROM PREDATORS

Burrowing predators can find their way to your flock if your

chicken coop doesn't have a floor or if it has a weak and rotted one. Snakes can get through very small holes, and you will likely not know they have entered your coop until it is too late. They will kill your chicken and eat their eggs. If you live in an area with snakes, elevate your chicken coop.

Rodents can't nest underneath the coop

Rodents usually nest in dark places. They love the safety of dark and protected areas. The rodents will make this place home if your coop has small spaces. If your coop is elevated, there will be too much space for rodents to hide as they will be easily exposed.

LOWER MOISTURE LEVELS

Building your coop on the ground could cause moisture problems, especially if you live in areas with seasonal floods, runoffs, or clay that doesn't drain well. If the floor is on the ground, it could be ruined by moisture or bring dangerous bacteria and mold into the coop.

THE COOP DOOR WILL NOT BE SNOWED IN

If you live in an area that experiences snowy seasons, a coop on the ground could have its entrance snowed in. You will have to shovel to clear it, which requires energy and time.

Boosts the durability of the coop floor

An elevated floor gets less wet, and a dry floor lasts longer.

Creates extra space for the chicken run

Your flock gets more room and can relax under the coop and use it as a shade or shelter from harsh weather conditions.

WAYS TO PREVENT PREDATORS

If you have chickens, you understand how important it is to protect them. Losing a pet can be very painful, so we are here to prevent this. Here is how:

KNOW YOUR ENEMY

Survey the area and identify the possible predators. It could be anything from hawks, raccoons and possums to owls. Once you have the answer, invest in prevention, and be aware that predators are very smart and can strike when you least expect it.

ENSURE THE COOP IS COVERED

This is a necessity in areas with air predators. You should cover all the areas your chickens roam around, especially the run. You may decide to use chicken wire, it's a good option as it still provides your chickens with visibility, but air predators can't attack them.

BURY THE CHICKEN WIRE

Predators will try to dig the ground to reach your chickens. When building a chicken run, bury the wire as deep as possible. Use a chicken wire to keep your flock safe and a mesh on the outside to keep predators out and away from your girls.

REPAIR THE ACCESS HOLES

Inspect the coop to determine if there are any access holes and fix them. You do not want snakes in your chicken coop.

Do a regular check for any signs of attempted entry and if there are any vulnerable areas, fix them immediately. You could line the coop or fence using metal siding.

LOCK YOUR COOP AT NIGHT

Ensure your coop is properly locked at night to prevent animals or humans from reaching your flock. Humans can also be a danger to your chickens, especially if you have a special breed or desired food. A raccoon is known to be a very intelligent animal and can open simple locks. Make sure you use a proper lock system.

CHECK BIOSECURITY

Cleaning your chicken coop is crucial. If you leave leftovers in the coop, it could attract rats which could eventually start feeding on your chicken eggs and chicks. They could even turn the coop into a home if you don't check it often.

COLLECT EGGS DAILY

Rats and snakes mostly get into your coop looking for eggs. If you collect the eggs daily, it will deter them from coming to your chicken coop.

GET A GUARD DOG

Predators cannot stand the scent of a dog, and they will likely decide to stay away from your flock. Your dog should be trained and well-sensitized to your chickens so that it doesn't become a danger to them.

INSTALL MOTION SENSOR LIGHTING

Lighting will deter a lot of predators, for example, raccoons which are very intelligent predators. They know when to attack, and they do it in the dark. When it approaches the coop, the alarm will detect movement, and lights will go on. The raccoon will be unable to attack.

INCREASE VISIBILITY

If you have a big garden, make sure you clear any potential hiding places for predators. The clearer the garden, the easier it is to see a predator approaching.

IS A BROODER NECESSARY?

A brooder is a heated enclosure that is used for raising chicks. Chicks rely on warmth for survival and can die if the mother or the brooder are unavailable. Brooders are also very important in hatching eggs. If you have one, you can hatch the eggs year after year. If you are new to chicks and don't have a brooder, let us help you identify some of the ones on the market.

UNDERSTAND YOUR BROODER
Chicken brooder box

This is a large container made to keep chicks in a particular space. If you want to bring your chicks home, this is what you are going to start with to raise your chicks. It fits small chicks perfectly as they do not require a lot of space. Ensure the brooder is twelve inches tall to prevent any sort of escape. Remember, baby chicks find it difficult to grip with their feet, so help them by using paper towels as bedding.

AREA BROODER

When your chick grows a little bigger, they are moved to this brooder from a brooder box. The area brooder is usually bigger than the box, so your chicks can enjoy more space. You can keep the chicks here until they are ready to move in with the rest of the flock.

HEATERS
Brooder heat plate

The heating plate is the most common of all the brooder heaters. It is also safe and easy to use. You only need to set it and introduce the chicks to it. With this brooder heating plate, you do not have to worry about fires as it has minimal risks of starting a fire. These heat plates maintain a constant temperature, so you do not have to worry about too much or too little heat.

BROODER LAMP

With this lamp, you only have to hang it on a chain, add another securing method, and then put it in the brooder. The temperatures of this lamp are not as constant as those of the heat plate, and you have to manually adjust it by lifting or lowering the lamp. Use a thermometer to determine the temperature and adjust the lamp's height accordingly. Unlike the brooder heat plate, a lamp poses a higher fire risk.

INFRARED HEATER

These heaters are just like the heat plate. They also reduce aggression in chicks. They are huge and distribute heat evenly; therefore, chicks will not fight for a specific spot as the whole area is warm. They are cost-effective and more efficient as they are thermostatically controlled.

WHAT SHOULD A BROODER HAVE?

- A heat source. You can pick one that suits you from the list above.
- A commercial or homemade perimeter.
- Maximum security from predators.
- A feeder and food. Get the right food for your chicks.
- An automated waterer or drinker and clean water.
- Suitable bedding for your chicks.

TYPES AND BENEFITS OF VENTILATION SYSTEMS

Fresh air is necessary to sustain your flock's life. It helps reduce air contamination, high temperatures, and humidity. If your coop is not ventilated, the composition of the air will change. Carbon dioxide, ammonia, and other dangerous gases will fill the coop. Ventilation is also used as a way of removing excess moisture as it prevents condensation on the walls and ceiling. Insulation affects heat and ventilation requirements. It reduces heat losses or gains through the wall and ceiling and controls condensation. The effectiveness of an insulator is measured by its R-value. The

higher the R-value, the more effective the insulation is. The amount of insulation in your coop will depend on the fuel cost and the area's climate conditions.

TYPES OF VENTILATION SYSTEMS

They are divided into natural air flow systems and mechanical air movement. Sometimes these are combined.

NATURAL AIRFLOW SYSTEM

A natural air flow system refers to external airflow into the coop due to pressure from natural forces. It should have an adequate air supply and distribution system to be called a good ventilation system. Air availability is controlled by the direction of the wind, orientation of the building, and site features. You should note that there are two types of natural airflow systems, wind-driven and buoyancy-driven ventilation. The wind-driven ventilation occurs when the pressure is created by the wind around a given structure or building, and openings form on the structure allowing airflow into the building. Buoyancy-driven ventilation happens due to directional buoyancy forces resulting from temperature differences between the interior and exterior. When there is a difference in temperatures between two volumes of air, the warmer air will be less dense, rising above the cold air, creating an upward air stream. For your chicken coop to be properly ventilated by use of this method, the temperature of the interior and exterior of the coop should be different. Wind-driven ventilation can further be classified as single-sided and cross-ventilation. Having an idea of the climate in your area is vital for deciding what kind of system to use.

BENEFITS OF BUOYANCY-DRIVEN VENTILATION

- It doesn't rely on the wind to function.
- It has a stable air flow compared to wind-driven ventilation.
- It is a sustainable method.
- You get to choose the areas of air intake.

DISADVANTAGES

- Buoyancy-driven ventilation has design restrictions and involves extra costs.
- Has a lower magnitude compared to wind-driven ventilation during windy days;
- It relies on different temperatures. The interior and exterior temperatures have to be different.
- The air introduced in the room could be polluted, especially if you are near an urban area.

MECHANICAL VENTILATION SYSTEM

This system ventilates a room in all extreme climatic conditions. An electric fan is used as the primary component for ventilation. Mechanical systems have four major components, the fan, heaters, openings, and controls. Fans and openings control air exchange in the coop. These fans are classified into two categories, negative pressure and positive pressure. The heaters supplement heat to maintain a desired temperature in the coop, while controls help adjust the ventilating, heating, and air velocity rates.

In the negative system, the fans are arranged in a way so that they expel air in the coop, giving room for fresh air to get in. The size and location of the fan are crucial in this system. The location of the fan will depend on the width of the building. If the building is about forty feet wide, you can place the fans on one side. The size of the air inlet is also critical for proper air ventilation. Its velocity should be high enough to ensure that fresh air gets into the entire coop, but also make sure that it is not too high as then it will expose your chickens to drafts. Install air inlets so that the air can move towards the ceiling. For a uniform air distribution, ensure the coop is airtight except for the air inlets, which should be left open. If there are any other openings, make sure they are fixed.

A positive pressure pushes air into a building and creates positive pressure. This system uses a fan to do so. The difference in pressure makes the air in the room leave through the outlets. One system pushes warm air into the coop and mixes it with the air in the coop, while the other pushes the warm air in the coop through plastic ducts with outlets. The whole system simply distributes heat and mixes air in the coop.

VENTILATION FACTORS TO CONSIDER

- Humidity, higher levels of humidity require more ventilation.

- The size of your flock, the bigger the number, the more ventilation you need.
- The type of breed you have, large breeds need more ventilation, as do those breeds that don't do well in hot areas.
- Seasonal changes, your coop needs to be ventilated for the whole year, but the amount of ventilation you need varies throughout the year. For example, you need more ventilation during summer and less during winter.

BENEFITS OF VENTILATION IN A COOP

Ventilation has many benefits for your chickens and is necessary for the survival of your flock. Without ventilation, high levels of humidity in the coop will put your flock at risk of suffering from various respiratory diseases. They can have breathing problems during any season, whether cold or warm. Here are some advantages of installing a ventilation system in your chicken coop.

- Eliminate harmful fumes. Chicken dropping releases a lot of ammonia which could fill the coop. Ammonia is not healthy for chickens. It causes eye irritation, respiratory problems, poor health, and even death. A properly installed ventilation system will ensure that the ammonia in the coop is expelled and not trapped inside. You can check if your ventilation system is functioning properly by determining if you can smell ammonia. If you can still smell it, improve your ventilation so that it functions as it should.
- Remove moisture from the coop. Even though your chickens don't sweat, they still produce moisture through breathing and droppings. This moisture makes the coop humid, which exposes the flock to respiratory problems. With a good ventilation system, you can get rid of the moisture in the coop.
- Provides fresh air. Your flock has a high respiratory rate, quickly depletes oxygen in the coop, and produces a lot of carbon dioxide. To ensure your flock survives, carbon dioxide has to be expelled to give room for oxygen. Ventilation does this best.
- Reduces the chances of airborne diseases by clearing dust and pathogens. The high respiratory rate of chickens puts them at a higher risk of getting airborne diseases. Without ventilation, the disease organisms will build up in the coop's stagnant air, making your flock ill. Ventilating the coop will push out stagnant air and allow fresh air to enter.
- Minimizes the heat produced during breathing. Heat is created by chickens when they breathe. Chickens do better in the cold and will not like it if the coop is too hot. If the temperatures go up to ninety degrees, your flock can die.

QUICK CHAPTER TAKEAWAY

In this chapter, you learn the different types of coops, how to build one at home, the necessary factors to consider before buying a coop, characteristics of a coop, elevation, and predation.

Chicken bedding, as mentioned earlier, is the new and clean material spread in and around a coop and in the nesting boxes. The purpose of having chicken bedding in your coop is to ensure that your flock is comfortable, healthy, and happy. The bedding reduces the amount of moisture and odor in the coop. It also ensures it's safe and comfortable for the chickens to play and jump around the coop. Ensure that the material you use for your flock is clean and easy to change. The rule is that you should change the bedding once every month. If you lose track and have no idea when you are supposed to change the bedding, you can always use your nose. If the coop smells, it is time to remove the old

bedding. Note that the bedding should be thick enough to provide ample cushioning.

Types of Chicken Bedding

The choice depends on your preference, but some bedding is more suitable for certain climates. You can use these materials to make bedding for your flock.

HEMP

Hemp can decrease the smell of ammonia from chicken poop. It is an absorbent material and provides warmth for your flock during the cold seasons. A hemp bedding is made of mulched cannabis stalked into a thick straw. Although it may appear tough, it is soft and, therefore, suitable to use in your nesting boxes. It is all-natural bedding and can be used as a natural insecticide. If you are allergic to other common bedding, hemp is a good option, although it can be expensive and difficult to find. Hemp can accumulate dust as well. Some advantages of using hemp are that they are highly absorbent, long-lasting, have good insulation, and act as a natural insecticide.

HAY

If you can't differentiate between hay and straw, you are not alone. Hay is a plant; straw is a byproduct of the grain crop. Hay is usually an all-time choice for many chicken keepers. It is cheap and absorbent. It also allows easy drainage of the chicken waste. Unfortunately, it isn't good at releasing moisture, and if you don't change this bedding weekly, it will be a breeding spot for bacteria and microbes and become smelly. Some advantages of hay bedding include its availability, low cost, good insulation, and ability to drain waste quickly.

PINE SHAVINGS

Pine shavings are a commonly used material because of their absorption capability and nice smell. It is readily available at an affordable price. Pine shavings dry quickly and last longer than hay, so it doesn't need to be changed as often. You only need to do it two to four times a year. Another important thing to note is that pine will not affect your flocks' respiratory system as much as cedar, but it will not insulate properly; therefore, it is not recommended for cold areas or during cold seasons. Some advantages of using pine include its fast drying, affordability, availability, and absorption.

STRAW

This traditional bedding has a natural smell and is a great insulator for your chicken coop, making it a good choice for cold seasons. You can get straw from barley, rye, and wheat. It's natural and, therefore, suitable for your chickens. Note that straw bedding has to be changed weekly to prevent any hazards. Some disadvantages of using straw bedding include that it molds easily, is not absorbent, requires frequent changing, and has low odor prevention. Advantages include great insulation, affordability, availability, it stimulates chickens mentally, and it is compostable.

SAND

Sand is a common choice. Understand that you cannot use just any type of sand as it could risk the health of your chickens. It is advisable to use mortar sand. It is the best for your chicken bedding as it does not rot or provide any breeding room for pathogens and bacteria. It also doesn't have odor issues. Sand is heavier than other bedding materials, so it's important to check if your chicken coop can withstand the weight. Avoid using sand during extreme weather conditions, as the sand could either get too hot or too cold. Some of the advantages of sand include that it: Helps with odor management, is easy to clean, lasts longer, and reduces bacteria and pathogens. Disadvantages are that it is very heavy, dusty, and provides poor insulation.

GRASS CLIPPINGS

Grass clippings are a very cheap option. They keep your flock occupied as they look for bugs and seeds in the grass. You can decide to do a deep litter method which will provide you with rich compost when the time to change the bedding comes. Advantages of grass clippings include that they are natural, free, and compostable, keep the chickens busy, and double as a treat. Disadvantages include the fact that they are only available seasonally, rot and smell readily, and offer poor insulation.

RECYCLED PAPER

These papers are commonly used as they are cheap and readily available. They maintain heat properly and, therefore, are a good choice for chickens in cold areas. Advantages include: they are good insulators, cheap, sustainable, and available everywhere. The disadvantages include: they require frequent changing, contain hazardous materials that are plastics and ink, do not reduce the odor in the coop, and can get slippery.

DEEP LITTER METHOD

A deep litter method is a housing system that involves the repeated spreading of the bedding material in indoor booths. You will spread an initial layer of the material, and when it starts going bad, you spread another new and clean layer on top of it. You will use six inches of organic bedding, which eventually turns into compost. This bedding does not require frequent changing. You can change it once or twice a year.

TRADITIONAL METHOD

In this method, you use light bedding that requires frequent changing. You can change it every day or every week, depending on the material you have used.

How Often Should You Change Your Chicken Bedding?

If you want a happy and healthy life for your chickens, you should ensure your chicken coop smells nice and fresh. You should do a simple clean daily and a thorough clean once weekly to get rid of the spoiled bedding and do a complete replacement. Your coop shouldn't get to a point where it smells like a health hazard for your flock. You can change your chicken bedding more often if you wish, especially if you have a larger flock. Although you can change the bedding as much as you want, it is still wise not to waste it. A daily spot clean with a complete weekly replacement will work well for your flock.

What are the Benefits of Chicken Bedding?

- The bedding is a cushion between your chickens and the floor and provides comfort.
- It acts as an insulator to shield the chickens from cold.
- Bedding promotes dryness in the coop.
- Absorbs moisture from chicken poop and drinkers.
- Enables the chickens to practice their natural behaviors with all the comfort they require.

- Clean bedding enables the chickens to have a restful sleep.
- Keeps the coop floor warm and clean.
- Provides a soft surface for your chickens to walk on.
- Bedding offers your chicken eggs a cushiony area to land, hence preventing cracks.
- It absorbs chicken poop and odor.

Understand What Diatomaceous Earth Is

It is a powder made from the sediment of fossilized algae found in water bodies and contains about 85 percent silica. This powder kills insects by dehydrating them. Although this powder dehydrates the parasites on your flock, it does no harm to the chickens. This powder allows liquids to flow through while capturing unwanted material. It is also used as a filler to prevent lump formation in foods, paints and plastics, pet litter, and medicine. It is used for various chemical tests and as an insecticide. In the past, diatomaceous earth was used to make building materials and later used in Europe for various industrial functions. When orally taken, it is used to supply silica which is key in treating high cholesterol levels, constipation, and problems with skin, hair, nails, and bones. Diatomaceous earth is mostly in a dust/ powdered form, although there are other forms the wettable powders and pressurized liquids. The products are permitted to use inside and outside your home and chicken coops. They kill bed bugs, spiders, fleas, ticks, and other pests that you can think of. Although diatomaceous earth has many health benefits, be careful not to inhale it. Breathing in this powder can cause nasal irritation, cough, or shortness of breath. It also poses a risk of eye irritation if it comes into contact with the eyes, and in some cases, skin contact can cause irritation.

TYPES OF DIATOMACEOUS EARTH

Although there are different types of diatomaceous earth, the base elements are the same. They are all made from processed diatoms. what makes it different is how these diatoms are processed. If you want to use diatomaceous earth for your flock, ensure it is 100 percent natural. Different varieties of the product are different in coarseness. You should use a fine variety for your chickens. The following are the types of diatomaceous earth.

FOOD GRADE

This type is suitable for consumption. It is purified and recognized as safe for humans and animals. It contains 0.5 to 2 percent silica which is used as an anti-caking agent and an insecticide in agricultural and food industries.

FILTER GRADE

Although this type has many industrial uses, such as dynamite production and water filtration, it is not safe to eat and is toxic to chickens. It is commonly known as a non-food grade. It contains more than 60 percent crystalline silica.

The Main Benefits of Diatomaceous Earth

- Kills pests. Diatomaceous earth kills both small and large insects. With this powder, you do not have to worry about lice, mites, fleas, or ants. Sprinkling the powder around your chicken coop will keep most pests away and kill those that stick around. Breeding will also be made impossible as the powder destroys their eggs.
- Diatomaceous earth neutralizes odor in the coop. It is great at absorbing bad smells. You can dust the powder in your coop, the bedding, and the nesting boxes. When using the deep litter method, avoid sprinkling the powder on it as it makes it rot quicker.
- It is a natural deworming remedy. This compound can be used to treat internal parasites in your chickens when mixed with the chicken feed or their drinking water. Be sure to use the food-grade diatomaceous earth, which is described above, so that it is safe for your flock. It will help kill parasites and help your flock grind down food like a normal grit.
- Acts as a grit and health supplement. In addition to its parasite-killing properties, it is rich in nutritional supplements for your flock. It is a source of essential minerals for improving your chickens' health. Undoubtedly diatomaceous earth improves egg quality and acts as grit. Feed your chickens 2-15 percent of the flocks' diet from this powder every day.

What Could Chicken Pecking Be?

Pecking is a natural bird behavior that allows them to check their surroundings, including their mates. Although pecking is a natural behavior, they may peck in an atypical manner if you leave your chickens confined for a prolonged period. Understand the kind of pecking that is going on with your chickens. Is it curious pecking or aggressive pecking? This will assist you in understanding if there is any problem amongst your flock. Aggressive pecking extends beyond what is considered normal in the hierarchy. It starts with feather pulling, especially in chicks, and could go on until the chicken gets injured to the extent of bleeding. The rest of the flock will be interested in the red color of the blood and could continue pecking it until they kill it. Aggressive pecking is a learned behavior, and the rest of the flock could adopt it. It can be challenging to eradicate this behavior, especially when it has been established in the chickens; therefore, you should implement preventative measures. The following are some measures: isolate any injured chickens until they have recovered. This will minimize the chances of the rest of the flock attacking it. Provide a balanced diet to your flock and ensure each chicken feeds daily. Provide mental stimulation for your flock and comfortable living conditions for your chickens, such as ensuring adequate space in the coop,

proper ventilation, and nesting boxes to ensure your flock is free from stress. Select docile breeds or separate docile chickens from aggressive chickens.

Chickens have a complicated social structure commonly known as the pecking order. This order is a social hierarchy that determines who gets to eat first and head the flock. Most of the time, it is usually the rooster in the flock. The rooster is naturally at the top of the order, although the pecking order will work differently if you have more than one rooster.

How Does Pecking Work in Single and Multiple Rooster Flocks?

If your flock has a single rooster, it will assume the position of the protector and provider, followed by the most dominant hen that will take the spot of a beta chicken, and then the rest will follow in the hierarchy. If there are multiple roosters, the setting changes, and the order starts with the roosters, although there is one who will be the head or, as commonly known, an alpha. He is usually in charge of all the other chickens. He is responsible for ensuring that the flock is fed and protected from danger. He is usually the first to crow, mate, and raise the alarm. This alpha will treat the hens kindly, and you will notice how he calls them with special squawks when he finds something for them to eat. He knows feeding them makes the hens trust him, making mating easier. Amongst the other roosters, one will become the beta rooster while the rest fall in line. The beta rooster works under the head rooster to ensure everything falls in line. Every chicken will know its position and role in the hierarchy and behave accordingly. A hen will take the alpha role if there is no rooster in your flock. After the alpha hen, there will be a beta hen that helps the alpha to lead. The beta hen occasionally challenges the alpha, but it gets pecked back into submission. The rest of the chickens will fall in line, although sometimes, they may try to challenge the alpha with no success. The henpecked chicken is the weakest of the flock. She gets bullied by the rest of the flock and prefers to stay away from them. She is the friendliest one, and most often, you will notice that it needs you to cuddle it when the rest leave to explore.

How to Stop Chicken Pecking

- Find out why your chickens are pecking. If your chickens start aggressive pecking, investigate the cause. Determine whether you've met the required standards for raising chickens. Make sure the chicken coop isn't overcrowded and properly ventilated, and that the flock gets enough water and food. Improve anything that may be stressing your chickens.
- Offer a place for your chicken to peck and keep your chickens busy. If they want to peck, provide them with an alternative, like a chicken toy, to prevent them from pecking each other.
- Keep your chickens clean. This will prevent feather picking.

HOW TO TAKE CARE OF A PECKED CHICKEN

Pecking is a natural habit in chickens from when they hatch and start exploring the world. You will see them using their beaks for almost everything. They peck each other, which can be disturbing if they are your first chickens. Note that they usually do this to regulate their social structure and not to cause harm. Sometimes chickens peck to cause harm, especially when they peck new chickens, chicks, sick chickens, and anything with blood on them. It is advisable to isolate an injured chicken, or the rest of the flock will kill it. You should only reintegrate the chicken when it has fully recovered with no signs of blood. Prepare yourself with a first aid kit as you don't know when this aggressive pecking

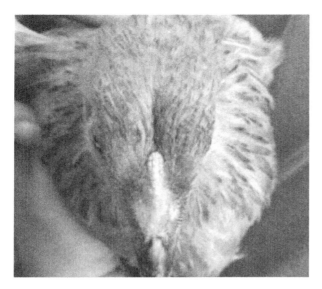

or predation will happen. Ensure the kit is well stocked. A chicken starts pecking whenever it gets stressed. There are three stages to saving an injured chicken, and they include:

- Isolation, cleaning, and examination.
- Healing and reintegration.
- Reducing aggressive pecking.

STAGE ONE

WATCH FOR PECKING AND INJURY SIGNS.

Observe your chicken occasionally, maybe twice daily, so you don't miss any signs of aggressive pecking. If you see any injury or a missing feather, assume it's aggressive pecking and isolate the chicken as soon as possible because if you don't, the flock will start pecking as they can see blood on the injured chicken. Also, learn how to differentiate between molting and pecking. With molting, the feather is usually crooked, but the feather is broken with pecking.

ISOLATE THE CHICKEN IMMEDIATELY.

It is crucial to isolate an injured chicken as soon as you can. It increases the chances of the chicken surviving the injury. It also lowers the chances of establishing an aggressive pecking behavior among the rest of the flock and the chances of cannibalism which is a behavior the flock can learn. Remember to isolate the chicken doing the severe pecking; you can integrate them when the injured one has recovered. When reintegrating them, start with the victim.

SET AN ISOLATION CAGE FOR THE INJURED ONE.

The isolation cage should be as comfortable as possible. Ensure there is good food, a drinker, a pecking toy, and nicely prepared bedding. The cage should be spacious enough and

properly ventilated. The attacking chicken should be isolated and out of sight of the injured chicken.

TEND TO THE CHICKEN TO STOP BLEEDING.

Put on clean gloves. It's time to tend to the wound. Place a clean cloth on the injury until the active bleeding stops. Hold the chicken safely in your arms and make it feel loved, calm it down as much as possible. At this point, the chicken is in pain and feels neglected. It is your responsibility to make sure it is calmer.

RINSE THE WOUNDED PART AND EXAMINE IT.

Hold your chicken and pour some warm water on its injury. Ensure you wash off all the dry blood before examining the wound and push the feathers away from the wound to have a clear view. If the wound is simply bruised, red, and has some bleeding, don't panic, you can nurse your chicken back to help without the help of a vet.

FOR ANY SEVERE AND INTERNAL INJURIES, SEEK A VET FOR HELP.

If you have tried to stop the bleeding without success, the wound covers a large area or is too deep, and you can't nurse it on your own, look for a vet. If the chicken can survive the injury, the vet will help. In cases where the vet cannot save the chicken, they may advise you on euthanasia.

STAGE TWO

Continue using wound spray.

After you have examined the wound, use poultry spray on

about mating for almost the whole year, and it increases during spring. You will see your roosters courting the hens by looking for morsels to impress the ladies, to increase their chances when mating time comes. Providing and caring for the hens will increase the chances of the ladies considering him when deciding who to mate with. Sometimes the hen will accept these privileges but will not mate with the rooster. They are picky and will mate with whichever rooster they feel comfortable around.

They can even refuse the alpha male if they don't like him. When a hen sees a rooster she does not like approaching, she will sit on her legs to show him that she has no interest in mating with the rooster, and she might even send it away. If another rooster tries to challenge him, he could be more submissive by chasing him away or starting a fight. The winner rooster will have the ladies and ensure that secondary roosters don't mate with his ladies. When the winning rooster gets old, which is around three years old; the younger and stronger roosters will challenge him and try to win over the ladies. Some roosters don't court. They go straight to business. Once he gets the 'opportunity,' the rooster will start treading to gain balance. He will grab the lady's comb and her head and neck feathers to balance himself further. It could appear violent, but it is usually the contrary. The rooster is usually gentle enough for the lady to tolerate. Sometimes there is an aggressive rooster who injures the ladies, and you should keep an eye on him as you may have to separate him.

it, which will help with recovery. You can use it three times a day or as the vet recommends. You should isolate the chickens at this stage.

CONCEAL THE WOUND

You can use a concealing spray to reintegrate your chicken and ensure the wound is covered. Concealing will discourage the rest of the flock from pecking.

PROVIDE SUPPLEMENTED WATER

To promote the healing of your bird, provide electrolyte-supplemented water. The chicken might not feed, so this shouldn't worry you. Just make sure that there is an ample supply of fresh water.

REINTEGRATE THE CHICKEN

When your girl is feeling better, you can try to reintegrate it; it's advisable to partially reintegrate it. For example, you can move its cage near the rest of the flock and watch how it gets on. If all is well, you can now fully integrate it. If it is attacked again, then permanently isolate it.

STAGE THREE

- Raise chickens that have reduced chances of aggressive pecking behavior.
- Create enough space for your chickens.
- Ventilate your chicken coop.
- Provide a balanced diet for your chickens.
- Provide pecking toys for your chickens.

Know How to Encourage Mating

Roosters ensure the continuity of their breed, they think

CHOOSE BREEDING FLOCK WISELY.

A chicken's health is crucial in breeding chicken, and so are its genetics. Before you breed your chicken, ensure there are no major genetic diseases in lines; you can do this by carrying out a breeding test and seeing what the initial stock produces. Some breeds have genetic disorders, which will carry out through the new generation of chicks. If your ladies are healthy, they will produce healthy babies. Note that if you want to improve the breed, ensure it mates with a chicken that meets the standard you are looking for.

SCHEDULE MATING DURING WINTER.

During hot seasons chickens spend most of their time trying to stay cool, and during winter, they preserve their energy to stay warm. This doesn't mean chickens don't produce fertilized eggs during this season. They do, although the process is more certain during spring. The roosters are more fertile during this period due to long days and warm temperatures. If you want to make mating easy for your flock, plan it for spring.

KEEP ROOSTERS AND HENS TOGETHER.

As we know, mating can't happen if the rooster and hen are kept separately. You should let the rooster have adequate time to impress and bond with the ladies. When they are kept together, the chance of mating increases. When the chickens are ready to mate, they will easily find each other. Make it easy for them.

GET THE RATIOS RIGHT.

A rooster requires five or more ladies to be contented. This doesn't mean you can't keep more than one rooster in your flock. If the flock becomes larger, you can increase the number of roosters. Adding roosters to your big flock will increase your chances of successful mating as a single rooster is unable to meet the flock's sperm production

demands, and it won't have the stamina to perform.

PROPER FEEDING.

If you feed your flock accordingly, they will have the energy to mate, it will also increase the rate of sperm production, and therefore the rooster will be able to fertilize a larger number of eggs. If your flock doesn't get food, they will spend time looking for food and won't have enough energy when it gets to mating time. Take care of your chickens so they have nothing else to worry about other than mating.

Common Problems During Mating.

- The roosters can be rough on the hen sometimes. Usually, when roosters injure the ladies, it results in minor damage like a broken or lost feather. You might notice your hens have a small, red, and irritated bald area or lost feathers. The rooster grabs the feathers, which leads to breakage. The rooster also grabs the back of the hen with his claws which leads to skin abrasions. Use a salve for mild cases and an antibiotic ointment as recommended. If the ladies are being injured a lot, you could use a saddle that will act as a barrier to protect the ladies' back from the rooster's talons.
- Rooster attacks during mating. Roosters don't mind sharing the same space if they each have a lady of their own. Hens usually don't cause issues as they are loyal and rarely change roosters. The problem comes in when there are competing roosters. There will be a fight when a rooster sees another rooster mating with a lady he wants. They will inflict wounds on each other, especially the combs, which can rip, causing a lot of bleeding. These attacks happen when a limited number of ladies are in the flock.
- Fertilizing eggs. It can be difficult to fertilize chicken eggs, especially the fluffier breeds. You will have to trim the bum fluff to ensure the cloaca is accessible by the rooster. This trim interferes with the hen's appearance.

WHAT ARE THE DIFFERENT CHICKEN BREEDING METHODS?

Chicken breeding can be more challenging than it appears. To avoid complications and undesired results, you should understand and plan for a method that will suit you. Knowing how to breed will help you improve the quality of the chicken products, create a source of income, and ensure you reach your goals for breeding. Knowledge is key in chicken breeding. Before you start breeding, have an action plan; understand why you need to breed and if you have the time to be with your flock through the process. Here are some methods you can use to breed your chicken:

FLOCK BREEDING

This is the most common chicken breeding method. It allows one rooster to breed with a couple of hens at random. It is easy for chicken breeders as you only have to put a large number of ladies with few cocks, preferably ten ladies to one cock. This

reduces operating costs. Some disadvantages of this method are: Multiple roosters tend to fight each other, male dominancy causes low fertility, and it doesn't maintain the pedigree.

PEDIGREE BREEDING

This is a method that involves breeding chickens to obtain specific characteristics. This method aims to mate a rooster to a typical lady to obtain the desired traits. You should take notes if using more than one lady when breeding for a show with this method.

INBREEDING

This method involves breeding related chickens; it improves good gene qualities and reduces the introduction of new genetic code problems. Some chicken genetic problems are recessive, and therefore if you breed them, these genetic problems could be dominant in the offspring. This is a great method if you want to breed quality chickens. You can predict how the chick will look depending on the parents' appearance. If you use this method for long, fertility levels will decrease.

LINE BREEDING

This method involves breeding within an ancestral line, for example, producing offspring from the father to daughter and mother to son. This is a safer method compared to sibling breeding.

OUTCROSSING

In this breeding method, you introduce new genetics but of the same line to an established line. Use this method if you want to correct sub-standard traits of a flock, like feathers or body issues. You should be careful when using this method, as new genetic codes could bring about complications.

CROSSBREEDING

This method involves breeding two chickens of different breeds. Like the other methods, it is used to correct genetic problems within a breed.

WHAT ARE THE ADVANTAGES OF BREEDING?

- For fresh meat and eggs, chickens are divided into three categories, breeding for meat, show birds, and egg production. Having fresh meat and eggs is a joy for any breeder. A single hen can produce 300 eggs a year, and those bred for meat can produce excellent quality meat if you take care of them as required.
- Backyard entertainment, all chicken breeders claim that watching their chicken in their backyard is entertaining. They enjoy every aspect of caring for the flock, even helping the baby chicken establish into their brooder. Chickens have different characteristics, and the breeder enjoys learning all these traits.
- Pets with benefits, chickens can help you control insects in your yard and boost soil health through tilling with their feet and beaks and naturally fertilizing the ground.
- Show chicken events. Chickens have other benefits apart from meat and egg production. You can use them for exhibitions during events.

QUICK CHAPTER TAKEAWAY

In this chapter, you will learn about the benefits of chicken bedding, the various types of chicken bedding, the basics of diatomaceous earth, mating, chicken pecking, and how you should care for your injured chicken.

These are commercial or homemade boxes for chickens to lay and hatch their eggs. A nesting box can only fit one chicken at a time. If you plan on keeping chickens for eggs, you will need these boxes; they create comfort and a secure area to lay eggs. You can purchase coops with nesting boxes or buy one without and purchase the nesting boxes separately. You also have the choice of building your nesting boxes. You can use several ideas and materials to build your nesting boxes, including food containers, milk crates, wood, dresser drawers, storage containers, totes, metal, cat litter tubs, feed buckets, and five-gallon buckets. Your nesting box should be:

- Safe, your flock should feel that it is safe to lay eggs in the nesting boxes. Invest in protecting these boxes from harsh elements and predators.
- Private and quiet, the best place to put your chicken nesting boxes is away from any traffic or noise. If the box location is too noisy, your chickens will lay eggs elsewhere or postpone laying, which is not ideal. When laying eggs, your ladies do not want to see each other. Therefore, it is important to keep it private.
- Dim lights, chickens prefer a dark place to lay eggs. If there are any lights in the coop, you can use curtains to cover the nesting boxes to reduce light.

These are some features you may want to include in your nesting boxes.

- Ventilation is key for raising chickens. You can achieve it by adding holes to the box using a solid container.
- Lips on the box to prevent the eggs and nesting material from falling out.
- An opening on the box for collecting eggs.
- Dried herbs are used to prevent odor, control insects, and provide soothing scents.

Types of Nesting Boxes Available

ROLL OUT NESTING BOX WITH CURTAIN

These boxes are either made of plastic or metal. These boxes have a slanted floor that enables eggs to roll away after being laid into the front or back of the box. It ensures that the eggs are safe from breakage and pecking and makes egg collection easier. You will collect high-quality and clean eggs.

WOODEN NESTING BOXES

You can find these boxes everywhere. They are easy to build, and you won't have trouble finding material. You can also easily get a commercially made one, but you must ensure it has the best material quality. A disadvantage of using this box is that you cannot put it outside or expose it to harsh conditions, like rain and sunlight, as it will rot.

METAL NESTING BOXES

These boxes are more steady, durable, and easy to clean. They are also light compared to wooden boxes and can withstand the wear and tear caused by the chickens. They are also available in different compartments, hence suitable for nesting chickens.

PLASTIC NESTING BOXES

Plastic boxes are cheaper compared to other types of nesting boxes and are suitable for first-time chicken keepers. They are sold separately; therefore, you can put them anywhere. They are also easy to clean.

BROWER 6 HOLE POULTRY NEST

This nesting box is made of galvanized steel and is thus durable. It is large and therefore provides comfort for your flock. The steel works as an insulator, adding warmth to the nesting box, which aids the hen in laying eggs.

FACTORS TO CONSIDER WHEN CHOOSING A NESTING BOX

The number of nesting boxes you require, the rule is one nesting box for every three or four hens. If you have a bigger flock, you could increase the number. At times you can have a dozen of nesting boxes, but the hens will still fight over their favorite nesting box. You can place some boxes in the coop and some inside the barn to cater to your flock's needs.

The size of your nesting box, size depends on the size of your flock. There should only be one hen in the nesting box to avoid breakage. Although it might seem great to have two hens in the box, avoid it and provide the privacy a hen needs during laying.

Your nesting boxes should be located in the coop or a private area. You should avoid putting the boxes near the feeding area or under the perches. You should raise them above the floor and put them in a dark area of the coop, and remember, roosts should be at a higher position than the boxes to ensure your ladies don't sleep in them.

The best material to use, the best material to use for a nesting box is either wood or plastic. They are easy to clean and durable. Wood is sturdy and has an attractive appearance, while plastic is light in weight, easy to clean, and durable.

Tips on how to take care of your chicken nesting box

Choose the location wisely. Different chickens have different preferences in nesting box locations. Still, avoid placing your nesting boxes in direct sunlight and being exposed to predators.

Clean the nesting box, especially after a chicken has taken a poop. Make sure when cleaning you change the padding material if necessary.

Choose the right nesting box for the breed. You should get a nesting box that suits your chicken breed. If it's a large chicken, the nesting box should fit their size.

Use padding in your nesting box; you can use a material of your choice to pad your nesting box. This will prevent egg breakage and comfort your hen when laying eggs.

Train the chickens on how to use the nesting boxes. Your flock will not know how to use a nesting box if you don't train it. It will continue laying in the bushes, giving you a hard time looking for eggs. You could purchase dummy eggs and place them in the box, letting the hen understand that that's where it should lay its eggs. When it continually sees eggs in the box, it will associate the nesting box with eggs and start laying there.

Use herbs in your nesting box. Herbs and insect repellants will discourage insects from entering the nesting boxes and release a good scent.

Create ventilation and drainage holes in the nesting box and if you are using a wooden box, avoid treating it with preservatives that could be toxic. The drainage holes should be installed at the bottom and ventilation holes at the top of the box.

Why Nesting Boxes?

- Reduces egg breakage. Nesting boxes keep your eggs safe so that they are whole when you collect them. A roll-away box will work best to prevent eggs from breaking.
- Easy to clean, nesting boxes are designed for easy cleaning. Cleaning your chicken nesting boxes is essential for the health of your flock. Dirty nesting boxes could affect the egg-laying capability of the hens. An easy-to-

clean coop will also reduce the rate of egg breakage.

- Nesting boxes are durable. They are made of strong material and are easy to assemble. They can withstand harsh weather and all the chicken activities of the day. The more you invest in a durable nesting box, the less often you will have to replace it.
- Provide a safe area for your chickens to lay eggs. Chickens can lay eggs anywhere as long as they feel safe. However, letting your chickens lay eggs just anywhere is risking predation. Your hen could go into the bushes to lay an egg and get attacked, injured, or killed.
- Saves you time and energy. Without a nesting box, you will have to walk around your backyard looking for areas where the hen could have laid and left an egg. This requires time and energy. With a nesting box, you only have to walk straight to your chicken coop and collect the eggs. You don't want to waste your morning energy and time looking for eggs instead of preparing a nice breakfast for your family.
- They act as insulators when your chickens lay eggs. Hens must be shielded from cold when laying eggs and nesting boxes meet this need.

Did You Know a Hen Can Lay Eggs Without A Rooster?

Your hen will lay eggs with or without a male. Without a male, the eggs are unfertilized and won't contain chicks that hatch. If the eggs are fertilized, remember to collect them every day, especially in the mornings, and put them in a cold place to prevent chicks from developing. If you want to own a rooster to breed with your chickens, remember some consequences come with that, the number of roosters in your flock will increase, and they could end up fighting from time to time which could lead to serious injuries. When they are still chicks, they can share a space but must be housed in different coops when they get to a certain age. Here are some of the considerations for keeping a rooster.

Firstly, check with the local authorities. Some areas or neighborhoods do not allow keeping roosters because of the crowing sound they make, especially in the morning. If your local authorities permit rooster keeping and you think the morning crowing could affect you or the neighbors, you could consider getting a night box. It restricts light penetration and movement. Your rooster will not be able to stretch his neck and crow, and it will not realize when it dawn, reducing the noise. You can let it free during the day. If you do not want to use a night box, you can use a rooster collar but ensure you keep an eye on him for any signs of breathing difficulty or distress. Remember that using a night box or rooster collar prevents the rooster from experiencing their naturally motivated traits. The other thing you should consider is comfort. Ensure the rooster has a clean and

comfortable coop, good nutrition, and other requirements that encourage natural behaviors.

Reasons why Some Chickens Do Not Lay Eggs

Chickens can stop laying eggs for several reasons. Some are natural responses; others can be fixed by making some adjustments. Make sure you confirm that your girls are not hiding their eggs before taking action. Let's look at some reasons that prevent your chicken from laying.

DAYLIGHT

Fewer light hours reduce egg production; your chickens need at least sixteen hours of daylight for sustained and strong egg production. Without light, your chickens will stop laying eggs due to hormonal responses as light time gets shortened. They will lay best if you expose your chickens to light for the recommended hours. If you are wondering why your chickens aren't laying as expected, check whether they are getting enough light.

WEATHER AND SEASONAL LAYING

Extreme weather does not work well with egg laying. When it's too cold or too hot, the hen cannot lay. Many ladies will not lay eggs in winter as there are fewer daylight hours, and the chicken is using a lot of energy to keep her body warm. The opposite is true during summer, but the result is the same- no eggs.

THE ENVIRONMENT OF YOUR COOP

If your hens get stressed, egg production will go down. Several factors can cause stress to your chicken, including too much noise, excess heat and cold, predators, aggressive mates, poor nutrition, and overcrowding. Assess to determine whether there are any stressors in the area.

WHEN SHE IS BROODY

A hen starts to sit on the eggs to hatch, so it stops laying eggs like usual. If you try to collect the eggs, she will still not lay eggs as long as she is still broody. Being broody is a hormonal thing, and therefore the hen can't just change her mind.

NUTRITION

Over-supplementing and over-treating your chickens will reduce egg production. These treats dilute the nutrients in the chicken feed. Therefore, the chicken cannot produce eggs consistently. For your girls to produce eggs consistently, they need thirty-eight nutrients in their bodies, calcium being the most important for egg laying. A complete layer feed is ideal for layers as it contains all the necessary nutrients in the right amount.

ILLNESS

When a hen falls ill, she stops laying eggs. She uses her energy to fight the illness and improve, reducing egg production. Your chicken will get better eventually and start laying as normal. Give it time.

TRAUMA

An injury or trauma will decrease the rate of laying eggs. For example, if a hen has a splinter, it will stop laying eggs until it recovers. The hen will get back on track as soon as it's healed.

PARASITES

Internal and external parasites can make a hen stop laying eggs. Check your ladies for parasites. Often there are parasite-like mites that cause anemia as they feed on blood, and worms decrease nutrient absorption. Find a solution for your chickens as soon as you can.

MOLTING

A chicken goes through molting annually. This starts when the chicken hits eighteen months and goes on until it dies.

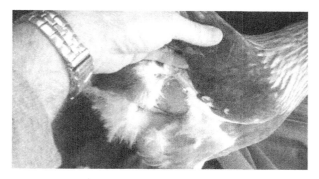

This is a season of feather loss and re-growth. This period is associated with a decrease in egg production. The hen's energy for egg laying is redirected to growing feathers. You shouldn't worry about reduced egg production for around eight to sixteen weeks. Give your girl time to recover, which will eventually return to normal.

THE AGE OF THE HEN

Some chickens may look full grown but are still babies. A hen can start laying as soon as they are eight months, but that depends on the breed. Some breeds, such as the Silkies, can start laying at nine months, while others start at four months. Ensure your chicken is fully grown before questioning why it's not laying eggs. Also, note that when a hen advances in age, egg production slows down, this happens when the hen is around two years.

Learn How to Help Your Chicken Lay Eggs During the Cold Season

Chickens lay eggs during winter as long as they live in favorable conditions and are the right breed. Let us look at what can help your chicken during this period.

THOROUGHLY CLEAN THE COOP.

You should spare some time to clean the coop to ensure your flock is comfortable. Scrub off all the dirt in the coop, run, nesting boxes, and roosting spots. Your chicken needs a clean and comfortable place without any stressors for the cold season. Your flock is going to spend most time indoors, and being in a dirty place for a long time will be difficult for them. They could end up not laying eggs.

PROTECT YOUR FLOCK FROM PREDATORS

Inspect your chicken coop for broken and vulnerable areas that can give the predators access. When your chickens

sense any danger, they stop laying eggs; therefore, fix what needs to be fixed to ensure the chickens are away from predators and harsh weather.

INSULATE THE COOP

Chickens, at this point, are struggling to keep their warmth. You can add an extra layer of bedding to keep the coop warm and stop the cold air from getting in. Remember, straw bedding is better than shavings or sand.

PROVIDE A HEATED WATERER

If you want your chicken to start laying, you'll have to provide a heated waterer or have a built-in metal cookie tin, with a kind of Christmas lighting, and hang it over the drinkers to provide heat and keep the water defrosted.

INSTALL A WALL HEATER

Wall heaters prevent frostbite. If temperatures are freezing, simply mount a ceramic wall heater to keep your flock warm.

HAVE LIGHTING ON A TIMER

When the coop is dark, your flock will spend more time sleeping than laying eggs. Therefore, you can set a light to a timer, and it will turn on early enough, that is, before dawn. Therefore, your hens will continue laying eggs during the cold season.

SPACING IN THE COOP

More chickens produce more heat; therefore, you shouldn't isolate a bird this season unless necessary.

Types of Chicken Eggs

There are different chicken breeds, and in the same way, there are different eggs. The eggs come in different colors, as mentioned earlier. The following are some of the colors available in the market. White, green, brown, blue, pink, and chocolate brown. The most common eggs are white and brown.

- Standard white eggs come from white chickens raised in conventional housing systems. These houses have been standard for many years.
- Standard brown eggs are from brown chickens raised in conventional systems.
- Enriched eggs are from chickens raised in furnished housing. These systems provide enough space to move around while they are being provided with a variety of enrichments. This enables the hens to express their natural behaviors.
- Free-run eggs are from chickens raised in a free-run type of system. The hens roam around as much as they want within the barn while providing enrichments, such as perches and nesting boxes.
- Free-range eggs are from hens raised in a barn or aviary (free-run) system. This system provides access to outdoor runs.
- Organic eggs are from chickens raised in a free-range system. The chickens are fed with certified organic foods only.
- Omega-3 eggs are from chickens fed food containing extra flax; therefore, the eggs contain omega-3 fatty acids.
- Vitamin-Enhanced eggs are from chickens fed with nutritionally enhanced foods with high levels of vitamins.

Therefore, the eggs contain high amounts of vitamins.
- Vegetarian eggs are produced by chickens that feed on plant-based ingredients only.
- Processed eggs contain added ingredients such as coloring and flavor.

Help Your Hen Say Goodbye to Eating Eggs

GET A SLANTED NESTING BOX

Purchasing or building a slanted nesting box like a roll-out nesting box will allow the egg to roll away after it is hatched, denying the chicken the opportunity to eat it or break it.

KEEP YOUR CHICKEN BUSY.

When your chicken is idle, it will start pecking on anything, even its own eggs. This is how they will start eating the eggs. To avoid this, find your chicken a toy to keep it busy. It won't remember the egg it laid.

COLLECT EGGS DAILY

Collect the eggs in the morning and do it twice or thrice daily. Leaving eggs for long will increase the chances of the chickens eating the eggs or breaking them.

PROVIDE YOUR CHICKENS WITH ENOUGH PROTEINS

Feed your chicken with enough protein, and ensure the protein level required for each stage of chicken raising is met. If you don't do this, the chickens will start eating the eggs in search of proteins. You could decide to supplement your flock's diet with sunflower seeds, milk, and yogurt.

KEEP EGGSHELLS STRONG

Feed your chicken with calcium for them to produce a strong shell. Strong shells are not easily broken or eaten. When a chicken realizes it can't break the shell, it will give up and walk away. A thin shell means it's either a broken or eaten egg.

PUT A GOLF BALL IN THE BOX.

The chickens will try to break the ball to no avail. They will understand that breaking an egg is an impossible task, and they will stop trying. This will save your eggs from being eaten.

PUT ENGLISH MUSTARD IN AN EMPTY EGG.

Many chickens don't like mustard, so you can trick them into believing that an egg is a big pot of mustard. You could fill an empty egg with mustard and leave it in the nesting box. The egg eater will be excited to find an egg, only to find a nasty surprise that will put him/her off from eating eggs.

PROVIDE A COMFORTABLE NESTING BOX

This will prevent breakage. If the hen lays an egg, it will fall softly on the bedding; therefore, without breakage, the chicken won't be curious to taste the egg, which could lead to egg eating.

KEEP NESTING BOXES DARK.

You can achieve this by installing nesting box curtains. After laying the egg, it will be difficult for the chickens to find the egg.

FEED CHICKENS COOKED, NOT RAW EGGS.

If you are considering supplementing your chickens' feed with eggs, don't give them raw eggs. They shouldn't have

a taste for raw eggs, or they might end up eating the eggs immediately after they have been laid. Cook the eggs you feed your flock.

INCUBATION AND HATCHING DISCUSSED.

Incubation is when a chicken keeps eggs warm for several days (21 days) until the young ones come. When the chick comes out, it is called hatching. The period of incubation depends on when the eggs are laid. When the hen starts incubating the eggs, her motherly instinct takes over. The hen does not go far away from the eggs. It is aware that if it neglects the eggs, her chicks might not develop properly or not hatch at all. You may have realized that modern chickens don't care about incubation, they are easily distracted by the rest of the flock, and that's why most chicken farmers have taken the responsibility of incubating into their own hands. How do you, as a chicken keeper, incubate your eggs?

Set up an incubator. This will depend on the number of eggs you want to incubate. With a good incubator, you only need to put the eggs inside, shut the door, and after three weeks, you have your chicks. You must ensure the temperature is set at 99.5 degrees during this period. If you change the temperature for some time, you will terminate the embryo. The incubator's humidity should be around forty to fifty percent for the first eighteen days and sixty five to seventy five percent for the last hatching period. Your incubator should also have vents that will allow fresh air circulation.

Find fertile eggs. After you've set up the incubator, it's time to look for fertile eggs. If you have a rooster, that means most of the eggs are fertile. You can pick them up and put them in the incubator. If you don't have any eggs, you can purchase fertilized eggs from a neighbor who raises chickens or a reliable supplier you know.

Incubate. Once you've got the eggs, it's time to incubate. Ensure you turn on your heat source and humidity while adjusting for about 24 hours to get the optimal temperature. When you are ready with the settings, put your eggs inside the incubator and wait for them to hatch.

A few days before hatching, you will see the eggs shift independently. This means that the fetus is active. The chick will peck the egg, break it, and take its first breath. The chick will relax for a couple of hours to enable its lungs to adjust, and then it will continue hatching. Do not help the chick hatch; it should do it on its own. After it has hatched, give it time to dry in the incubator and move it to a brooder.

How to Raise Baby Chickens in the Comfort of Your Home

Taking care of adult chickens is easy, but raising chicks has a high mortality rate. Raising chicks depends on having all the supplies required and frequently monitoring them. You need to follow certain steps to ensure your chicks survive. Follow to ensure good health for your chicks.

- Setting up chick feeders.
- Ensure you fill the waterers and feeders.
- Set up your chick brooder.
- If you are getting the chicks from somewhere else, be ready to receive them. They shouldn't wait long as they need urgent care.
- Dip each chick's beak in water and ensure they get some drinking water.
- Observe the chicks' behavior to determine any abnormalities or needs you should meet as soon as possible.
- Check for pasting up. Check if there is any poop stuck around the feathers around the anal vent, which could hinder defecating.
- Always keep the bedding clean.
- Ensure you feed the chicks and provide for all their needs.
- When they are grown enough, you move them to the coop.

Chapter 9
Chicken Health Problems

Chickens also suffer from diseases. As a chicken keeper, you should minimize the chances of this happening. Here are some health issues or diseases that can affect your flock.

COCCIDIOSIS

It is a disease caused by a parasite commonly known as Coccidian protozoa. This parasite lives in the chicken, damaging a specific area in the chicken's gut. This disease begins when your flock takes in sporulated oocysts, which are broken down in the gut to release an infectious sporocyst, causing damage to intestinal epithelial cells.

FOWL CHOLERA

It is caused by Pasteurella multocida that affects joints, infraohits, sinuses, wattles and other tissues. It is mostly found in adult chickens and affects males more than females. Some symptoms include loss of appetite, swollen joints, diarrhea, and ruffled feathers.

FOWL POX

Avian Pox is another name for Fowl Pox. This is a highly contagious disease. There are two categories of this disease: Dry Pox and Wet Pox. The common symptom of this disease is a bump that looks like warts visible on the chicken's comb and wattles. Affected young chickens will have reduced egg production and stunted growth.

AVIAN INFLUENZA

It is caused by type A Orthomyxovirus. This disease is commonly found in wild birds. It has the following symptoms: nasal discharge, diarrhea, discoloration, coughing, and edema. Avian Influenza is a deadly disease.

SALMONELLOSIS

This is a bacterial disease that causes septicemia and enteritis in chicks. Infection is contracted orally, and rodents are likely to spread it.

NEWCASTLE DISEASE

This is a respiratory disease that spreads rapidly. Symptoms depend on whether the infecting virus strain prefers the respiratory, nervous, or digestive systems. It affects both wild and domestic animals, but domestic animals are more likely to get it.

How Do You Know Your Chicken is Having Health Problems?

Here are the symptoms that will help you identify a sick chicken.

- Feather loss
- General inactivity
- Abnormal droppings
- Discharges
- Dull and closed eyes
- Drooped wings
- Ruffled feathers
- Lying down

Ensure Your Flock Stays Healthy

Let us recap what you should do to ensure your flock stays healthy. I am positive that you have an idea from the previous chapters by now.

- Provide adequate space for your flock.
- Provide them with good food.
- Keep them dry.
- Protect them from harsh weather, aggressive chickens, diseases, and predators.
- Let your chicken exercise.
- Provide fresh water.
- Keep an eye on them and bond with them.
- Clean the coop for them.
- Provide proper bedding.
- Limit treats.
- Provide dust bath areas.

Can Your Flock Make You Ill?

The answer is yes. Your cute babies can make you sick with

Salmonella infections. These chickens can appear healthy while they are carrying these bacteria. To avoid infections, kindly follow the following steps. Always clean your hands well with soap and water after being in contact with your chickens. Wash your hands after touching the clothes that you wore while taking care of your flock. If your child was in contact with the flock, supervise them while they wash their hands. Do not allow chickens in your space, especially where you keep your food and drinks, and remember not to eat near or inside your chickens' houses. Anyone with weak immune systems, such as kids and older people, should not get into contact with the flock. Avoid kissing your chickens or allowing them to touch your mouth. Ensure you stay outside when cleaning equipment for raising chickens, and finally, go through the CDC's recommendations on taking care of your chickens.

To start raising chicken as a business, you should research and understand the appropriate sector and the market and demand. After you have this sorted, understand what your local authorities say about raising chicken businesses, if they permit raising chicken, and what restrictions exist. If the authorities permit raising chickens in your area, develop a business strategy. How are you going to start? How much do you need for the project, do you have all the documentation, and how will you advertise your products? If your plan is well laid out, it is possible to attract investors; they will see that you are determined and feel the urge to invest in your business, don't be shy about trying out investors to see if they are interested in your project. Do your math properly to determine the size of the flock you want to keep and how you want to keep it. You can find a name for your business, create a logo or even a website, and then create networks. Networking will boost your business, and you can start attending seminars and any other events on raising chickens. You will be able to learn one or two things about raising chickens and get insights from fellow farmers. These are some of the businesses that you can start.

EGG SALE BUSINESS

Research is key in deciding the product you are going to sell. If you have settled on eggs, you need to understand the market. An egg is a very important product, and most people want eggs. You will be able to enter the market easier if you have a good business plan, understand the types of eggs selling in your neighborhood, and how much an egg goes for. Who will you be selling these eggs to? These are questions you should have figured out before starting the business. If you want to sell 200 eggs in a week, determine the number of chickens you need to produce this, the price you will sell the eggs at, and the profit you will make. You do not want to start a business just for it to fail because you aren't making any profits.

CHICKEN BROILER BUSINESS

If you have decided to go with selling meat, you need to start with a business plan and state your goals and what you would like to achieve. Your achievement should have a timeline. Like with the egg business, determine whether there is a meat market. Consider the basics, which are capital, equipment, and land. Decide the breed you want to sell and how you want to sell them. If all goes well, increase the size of your flock. You can talk to people who understand the broiler business, and they will help you. For example, if you want to sell chicken, do you sell it alive or slaughtered, and how much do you sell it for? Are you going to open a shop, or will you sell online? If you get everything right, your business will be successful.

CHAPTER TAKEAWAY

You get to understand a lot in this chapter; it involves understanding the different health issues chickens face and how you identify a sick chicken. You also learn how to ensure your flock remains healthy and the business opportunities you can grab.

Conclusion

Chickens are very important in our lives, from when they are hatched and join our beautiful world. We watch them grow and withstand all their difficulties, diseases, harsh weather, and even losing some of their siblings. They remain strong and grow into adulthood and provide us with food. They provide eggs and meat for food and poop: which creates compost for sale or to use in your garden. Chickens enlighten our homes. Several families have raised chickens and kept them as pets. Even though sometimes they can be noisy, their presence warms our hearts. If you take care of your flock and ensure they are happy and healthy, they will always run to you when you return from work. Who doesn't want a warm welcome after a long day at work?

In our book, we have provided detailed information on raising chickens without leaving any information out. We are sure you now feel ready to start chicken keeping. We have started right from the beginning, helping you understand the history of chicken domestication as it's key to understanding chickens. We have also introduced you to key chicken terminology, so you won't be left behind. The benefits and what you require to get started are all provided in this book. We understand that raising chickens isn't easy for many, so we compiled this book to explain the many benefits, challenges, chickens' natural behavior, local restrictions, and how you can calculate your budget and do your research.

Our book also shows you the different breeds you can choose from, and which breeds suit you and your environment. Some breeds are better off in cold seasons, while others are better off in warm areas. You don't need to worry about what equipment you need. We have done the research and written it down for you in a way that is easy to understand. This book is all you need.

You will also find details on how to take care of sick or injured chickens in this book.

You can raise chickens for several reasons. Maybe it's your hobby, a business, a pet, or the fact that you want food for your family.

We hope this book brings you solutions to your problems. As we promised at the beginning of this book, we deliver. You can wave bye-bye to all the disappointments. Happy raising chickens.

BENJAMIN D. NELSON

Printed in Great Britain
by Amazon